ILLINOIS STATE UNIVERSITY LIBRARY

P9-CAO-995

3 1611 00193 9112

UNIVE
GOVERNORS STATE UNIVERSITY
PARK FOREST SOUTH, ILL.

DATE DUE

JAN 5 1984

The Competitive Status of the U.S. Auto Industry:

A Study of the Influences of Technology in Determining International Industrial Competitive Advantage

Prepared by the Automobile Panel,
Committee on Technology and
International Economic and Trade Issues

of the Office of the Foreign Secretary,
National Academy of Engineering

and the Commission on Engineering
and Technical Systems,
National Research Council

UNIVERSITY LIBRARY
GOVERNORS STATE UNIVERSITY
PARK FOREST SOUTH, ILL

NATIONAL ACADEMY PRESS
Washington, D.C. 1982

NOTICE: The project that is the subject of this report was approved by the Governing Board of the National Research Council, whose members are drawn from the Councils of the National Academy of Sciences, the National Academy of Engineering, and the Institute of Medicine. The members of the committee responsible for the report were chosen for their special competences and with regard for appropriate balance.

This report has been reviewed by a group other than the authors according to procedures approved by a Report Review Committee consisting of members of the National Academy of Sciences, the National Academy of Engineering, and the Institute of Medicine.

The National Research Council was established by the National Academy of Sciences in 1916 to associate the broad community of science and technology with the Academy's purposes of furthering knowledge and of advising the federal government. The Council operates in accordance with general policies determined by the Academy under the authority of its congressional charter of 1863, which establishes the Academy as a private, nonprofit, self-governing membership corporation. The Council has become the principal operating agency of both the National Academy of Sciences and the National Academy of Engineering in the conduct of their services to the government, the public, and the scientific and engineering communities. It is administered jointly by both Academies and the Institute of Medicine. The National Academy of Engineering and the Institute of Medicine were established in 1964 and 1970, respectively, under the charter of the National Academy of Sciences.

This project was supported under Master Agreement No. 79-02702 between the National Science Foundation and the National Academy of Sciences.

Library of Congress Cataloging in Publication Data

Main entry under title:

The competitive status of the U.S. auto industry.

Bibliography: p.
1. Automobile industry and trade--United States.
I. National Academy of Engineering. Committee on Technology and International Economic and Trade Issues. Automobile Panel. II. National Research Council (U.S.). Committee on Engineering and Technical Systems.
HD9710.U52C58 1982   338.4'762920973 82-12506
ISBN 0-309-03289-X

Available from

NATIONAL ACADEMY PRESS
2101 Constitution Avenue, N.W.
Washington, D.C. 20418

Printed in the United States of America

HD
9710
.U52
C58
1982
c. 1

Participants at Meetings of the Automobile Panel,
Committee on Technology and
International Economic and Trade Issues

Panel

WILLIAM J. ABERNATHY (Chairman), Professor, Harvard
University Graduate School of Business Administration
ALAN A. ALTSHULER, Professor and Chairman, Political
Science Department, Massachusetts Institute of Technology
JAMES K. BAKKEN, Vice-President, Operations Support
Staff, Ford Motor Company
DONALD F. EPHLIN, Vice-President and Director of National
Ford Department, United Auto Workers
DONALD A. HURTER, Manager, Automotive Technology Unit,
Arthur D. Little, Inc.
TREVOR O. JONES, Vice-President and General Manager,
Transportation and Electrical and Electronics Operations,
TRW, Inc.
HELEN R. KAHN, Bureau Chief--Washington, Automotive News
DUANE F. MILLER, Vice-President, Engineering, Volkswagen
of America, Inc.
RALPH L. MILLER, Vice-President of Market Development and
Strategic Planning for the Light Vehicles Group, Rockwell
International*
RICHARD H. SHACKSON, President, Shackson Associates,
Inc.†
PETER D. ZAGLIO, Vice-President--Securities Division,
Lehman Brothers Kuhn Loeb

---

*Formerly, Director, Manufacturing Facilities Planning, General
Motors Corporation.
†Formerly, Assistant Director of Transportation Programs,
Energy Productivity Center, Carnegie-Mellon University.

iii

Rapporteur

KIM B. CLARK, Assistant Professor, Harvard University
    Graduate School of Business Administration

Additional Participants

JOHN ALIC, Project Director of International Security of
    Commercial Programs, Office of Technology Assessment,
    U.S. Congress
LARRY BURDETT, Vice-President, Marketing, Gelco, Inc.
WILLIAM CHANDLER, Director of Energy Conservation
    Project, Environmental Policy Center
ROBERT COLEMAN, Automobile Industrial Specialist, U.S.
    Department of Commerce
PETER FROST, Office of Advanced Technology, U.S.
    Department of State
KEN FRIEDMAN, Director of Policy Coordination and
    Support, U.S. Department of Energy
RICHARD JOHN, Director, Office of Energy Environment,
    Transportation Systems Center, U.S. Department of
    Transportation
WILBERT JONES, General Engineer, Office of Industrial
    Policy, U.S. Department of Commerce
ALAN KANTROW, Associate Editor, Harvard Business Review
LUCY LAUTERBACH, Presidential Management Intern, Office
    of Industry Policy, U.S. Department of Transportation
LARRY LINDEN, Senior Policy Analyst, Office of Science
    and Technology Policy, Executive Office of the President
MARK MARTICH, Sloane Fellow, Massachusetts Institute of
    Technology
RICHARD NANTO, Analyst in International Trade & Finance,
    Congressional Research Service, Library of Congress
ROLF NORDLIE, Director of Transportation and Heavy
    Equipment Division, Office of Producer Goods, Bureau of
    Industrial Economics, U.S. Department of Commerce
SUMIYE OKUBO, Policy Analyst, Division of Policy Research
    and Analysis, Scientific, Technological, and International
    Affairs, National Science Foundation
ROLF PIEKARZ, Senior Policy Analyst, Division of Policy
    Research and Analysis, Scientific, Technological, and
    International Affairs, National Science Foundation
ALAN RAPOPORT, Policy Analyst, Division of Policy
    Research and Analysis, Scientific, Technological, and
    International Affairs, National Science Foundation

BRUCE RUBINGER, Operations Research Analyst,
    Transportation Systems Center, U.S Department of
    Transportation

Consultant

BENGT-ARNE VEDIN, Research Program Director, Business and
    Social Research Institute, Stockholm, Sweden

Staff

HUGH H. MILLER, Executive Director, Committee on
    Technology and International Economic and Trade Issues
MARLENE R. B. PHILLIPS, Study Director, Committee on
    Technology and International Economic and Trade Issues
ELSIE IHNAT, Secretary, Committee on Technology and
    International Economic and Trade Issues
STEPHANIE ZIERVOGEL, Secretary, Committee on Technology
    and International Economic and Trade Issues

UNIVERSITY LIBRARY
GOVERNORS STATE UNIVERSITY
PARK FOREST SOUTH, ILL.

v

Committee on Technology and
International Economic and Trade Issues (CTIETI)

Chairman

N. BRUCE HANNAY, National Academy of Engineering Foreign
Secretary and Vice-President, Research and Patents, Bell
Laboratories (retired)

Members

WILLIAM J. ABERNATHY, Professor, Harvard University
Graduate School of Business Administration and Chairman,
CTIETI Automobile Panel
JACK N. BEHRMAN, Luther Hodges Distinguished Professor of
International Business, University of North Carolina
CHARLES C. EDWARDS, President, Scripps Clinic and
Research Foundation and Chairman, CTIETI Pharmaceutical
Panel
W. DENNEY FREESTON, JR., Associate Dean, College of
Engineering, Georgia Institute of Technology and Chairman,
CTIETI Fibers, Textiles, and Apparel Panel
JERRIER A. HADDAD, Vice-President, Technical Personnel
Department, IBM Corporation (retired)
MILTON KATZ, Henry L. Stimson Professor of Law Emeritus,
Harvard Law School
RALPH LANDAU, Chairman of the Board and Chief Executive
Officer, Halcon International, Inc.
JOHN C. LINVILL, Professor, Department of Electrical
Engineering, Stanford University and Chairman, CTIETI
Electronics Panel
RAY McCLURE, Program Leader, Precision Engineering
Program, Lawrence Livermore Laboratory and Chairman,
CTIETI Machine Tools Panel

vii

BRUCE S. OLD, President, Bruce S. Old Associates, Inc.
and Chairman, CTIETI Ferrous Metals Panel
MARKLEY ROBERTS, Economist, AFL-CIO
LOWELL W. STEELE, Consultant--Technology Planning and
Management*
MONTE C. THRODAHL, Vice-President, Monsanto Company

---

*Formerly, Staff Executive, General Electric Company.

Preface

In August 1976 the Committee on Technology and International Economic and Trade Issues examined a number of technological issues and their relationship to the potential entrepreneurial vitality of the U.S. economy. The committee was concerned with the following:

• Technology and its effect on trade between the United States and the other countries of the Organization for Economic Cooperation and Development (OECD).
• Relationships between technological innovation and U.S. productivity and competitiveness in world trade; impacts of technology and trade on U.S. levels of employment.
• Effects of technology transfer on the development of the less-developed countries and the impact of this transfer on U.S. trade with these nations.
• Trade and technology exports in relation to U.S. national security.

In its 1978 report, Technology, Trade, and the U.S. Economy,* the committee concluded that the state of the nation's competitive position in world trade is a reflection of the health of the domestic economy. The committee stated that, as a consequence, the improvement of our position in international trade depends primarily on improvement of the domestic economy. The committee further concluded that one of the major factors affecting the health of our domestic economy is the state of industrial innovation. Considerable evidence was presented

---

*National Research Council, 1978. Technology, Trade, and the U.S. Economy. Report of a workshop held at Woods Hole, Massachusetts, August 22-31, 1976. National Academy of Sciences, Washington, D.C.

during the study to indicate that the innovation process in the United States is not as vigorous as it once was. The committee recommended that further work be undertaken to provide a more detailed examination of the U.S. government policies and practices that may bear on technological innovation.

The first phase of the study, based on the original recommendations, resulted in a series of published monographs that addressed government policies in the following areas:

- The International Technology Transfer Process.*
- The Impact of Regulation on Industrial Innovation.*
- The Impact of Tax and Financial Regulatory Policies on Industrial Innovation.*
- Antitrust, Uncertainty, and Technological Innovation.*

This report on the automobile industry is one of six industry-specific studies that were conducted as the second phase of work by this committee. Panels were formed by the committee to address electronics; ferrous metals; machine tools; pharmaceuticals; and fibers, textiles, and apparel. The objective of these studies was to (1) identify global shifts of industrial technological capacity on a sector-by-sector basis, (2) relate those shifts in international competitive industrial advantage to technological and other factors, and (3) assess future prospects for further technological change and industrial development.

As a part of the formal studies, each panel developed (1) a brief historical description of the industry, (2) an assessment of the dynamic changes that have been occurring and are anticipated as occurring in the next decade, and (3) a series of policy options and scenarios to describe alternative futures for the industry.

The methodology of the studies included a series of panel meetings involving discussions among (1) experts named to the panel, (2) invited experts from outside the panel who attended as resource persons, and (3) government agency and congressional representatives presenting current governmental views and summaries of current deliberations and oversight efforts.

The drafting work on this report was done by Kim B. Clark, Harvard University Graduate School of Business Administration. Professor Clark was responsible for providing research and resource assistance as well as producing a series of drafts, based on the panel deliberations, that were reviewed and critiqued by the panel members at each of their three meetings.

***

*Available from the National Academy of Engineering, Office of the Foreign Secretary, 2101 Constitution Avenue, N.W., Washington, D.C. 20418.

Contents

Appendixes

The Competitive Status of the U.S. Auto Industry

Summary

The U.S. automobile industry is in a crisis. Vigorous import competition, drastic shifts in consumer preferences, and anemic final sales combined to make 1980 and 1981 two of the most difficult years in the industry's history. The current picture is bleak: literally hundreds of thousands of people have lost their jobs; communities dependent on the industry have suffered devastating losses in employment and financial resources; all the domestic producers have suffered major financial losses; large facilities have permanently closed. Future prospects are uncertain. If the industry is to survive, the next five years will see wrenching changes in its productive and financial base as new product technologies are introduced, manufacturing plants are retooled; and new relations are established among management, labor, and government.

Given its size and scope, it is not surprising that the auto industry has long been accorded significant public attention. Recent events have prompted debate about the current crisis and appropriate courses of action. Three alternative lines of interpretation can be distinguished; as defined in Chapter 1, they can be summarized as follows:

Transient Economic Misfortune. Proponents of this view argue that while the current crisis is a serious misfortune, it is temporary. The essential problem is a lack of small-car capacity; its solution is sufficient time and money to realign the product line.

Natural Consequence of Maturity. Based on theories of the product life cycle, this view treats the current crisis as one episode in a long-term shift of production out of this country to lower-cost sources of supply.

Fundamental Structural Change. This view challenges the notion that technology is stable. It envisions a period of rapid innovation in products and processes, where competitive advantage will depend on the ability to innovate.

These three interpretations make different assumptions about the competitive cost (and quality) position of U.S. production and the role of technology in competition. Moreover, they have quite different implications for public and private policy. This report places the cost and technology issues in their historical context and examines evidence on the character of recent innovation and the U.S. auto industry's relative competitive position. It should be noted that the report makes no attempt to estimate the comparative advantage in U.S. automobile production in the sense of classical economic trade theory (e.g., the ratio of U.S. costs of auto production relative to U.S. costs of other goods, compared with similar ratios for other countries). Rather, the report examines the competitive position of the U.S. producers relative to their major competitors within the auto industry. The focus is not only on costs of production but also on product quality and the role of technology and innovation in competition.

THREE HISTORICAL THEMES

A basic premise of this report is that the nature of the current crisis in the automobile industry, the specific problems faced, the patterns of observed response, the barriers to adjustment, and the strengths and weaknesses of domestic firms can be understood only if one first understands something of the history of the industry. Chapters 3-5 of the report sketch out three themes that have characterized the evolution and development of the industry in the United States: the convergence of technology, the internationalization of markets, and the growth of public demand on the industry.

Convergence in Technology

Anyone trying to buy a car in 1905 was confronted by considerable variety: steam cars, electric cars, cars powered by gasoline, cars with three or four wheels, open-air cabs, closed carriages, all manner of mechanical principles. By 1973 that technological diversity had disappeared. To be sure, there was immense variation in styling, but the underlying technology had become standardized. This standardization of technology reflected a change in the character of innovation as well as a particular pattern of competition.

From an early stage in which technical change was rapid and fundamental, the industry evolved to a point where technical advance was incremental and almost invisible. Competition was oriented toward the mass market where cost, styling, and acceptable levels of performance were the basic dimensions of competition. In contrast to the European market where technology-based competition led to technological diversity, convergence in technology in the United States reflected a market where product technology was competitively neutral.

Internationalization

Until the last decade the evolution of the U.S. automobile industry was largely determined by political and economic forces specific to the North American continent. In the last few years, however, the relevant industry boundaries have expanded dramatically. There have been two interrelated changes: (1) dramatic shifts in the volume and pattern of world trade and (2) growth in the number of viable competitors worldwide.

As played out in the United States, internationalization occurred primarily in the small-car segment through import penetration; the U.S. response has been conditioned by a legacy of large-car production. Large cars were associated with luxury and prestige and commanded premium prices; in terms of cost, however, small cars were just about as expensive to make; the result: small cars, small profits. With little incentive, U.S. producers did not develop products to compete directly with the imports until the late 1970s. Internationalization has confronted U.S. producers with competitors operating with a very different competitive tradition and experience. It is now clear that success in small cars requires different capabilities than success in the large-car segment; attaining parity in subcompacts with foreign producers involves far more than realigning the product line.

Public Demands on the Industry

During the last 10 years the development of automotive competition and technology has been strongly influenced by government mandate. Social demands on the industry are not new; manufacturers have long had to meet both the demands of the marketplace and the requirements of changing social expectation.

From the early years of the industry up to World War II, market demands and public demands coincided. In the 1950s and 1960s, however, perceptions shifted and new public demands were imposed. Concern for safety, pollution, and energy efficiency led to a variety of government initiatives. The specific form that evolved--mandated standards and agency regulation--reflects public perception of the industry as a "bad guy" and the tension between divergent social and market demands.

In addition to specific government policy directed at the industry, the postwar era has demonstrated the impact that general policy measures can have. In recent years, for example, the sensitivity of the industry to general economic conditions and the stance of fiscal and monetary policies has been underscored by record high interest rates, sluggish economic growth, and falling real income. Coming at a time when the industry's need for resources to meet new competitive demands has been at an all-time high, the depressed state of the automobile market has dealt the industry a severe blow.

Competition in the U.S. auto industry has undergone fundamental changes in the last 10 years, primarily because of increased market penetration by foreign manufacturers and drastic shifts in the price of oil. The events of the 1970s confronted a mature industry used to competing on the basis of scale economies, styling, and dealer networks. It was an industry in which technology in particular and manufacturing in general had become competitively neutral. It was an industry increasingly subject to government mandates, competing on an international basis with new competitors who emphasized superior manufacturing performance. Moreover, growing incentives for new technology have created the opportunity, even the necessity, for competitive advantage through innovation.

As we noted at the beginning, there are wide disagreements about the meaning of the current crisis. The three categories of interpretation sketched out in Chapter 1 differ in the assumptions made about the relative costs and quality of U.S. products and the stability of technology. Chapters 6-9 present evidence that bears on these issues.

Product Cost and Quality

Our analysis of productivity and product cost makes use of a variety of sources of information, including government reports and other published analyses as well as studies conducted within companies in the industry and made available by members of the panel. (Where use has been made of internal company analyses, trip reports, or other "industry sources," these have been explicitly noted.)

Based on a variety of approaches and data sets, we find that the Japanese have a significant landed-cost advantage. Although differences in the two systems of production make precise comparisons difficult, the Japanese advantage is likely to fall in the range of $750 to $1500 per small vehicle. Evidence on the cost differences from publicly available information is presented in Appendix A. While the data presented there are consistent with the finding of a sizeable cost advantage for the Japanese, the precise order of magnitude and the confidence that industry members of the panel place in the cost difference (i.e., $1200-$1500) comes more from internal studies using confidential and proprietary data. The Japanese advantage reflects differences in prices as well as productivity. Compared with the U.S. firms, the major Japanese producers (Toyota, Nissan, etc.) have significantly higher overall productivity (total employee hours per vehicle); some estimates put the productivity difference as high as 40-50 percent. Employee cost per hour worked in Japan is about 50-60 percent of the U.S. average.

The analysis of cost and productivity has implications for comparisons of profitability between U.S. and Japanese auto companies. Because the Japanese firms sell their cars in the United States at prices that are comparable with prices for U.S. cars, the cost advantage of the Japanese gives them a higher margin of profit on cars sold in the United States than that of the U.S. manufacturers. Evidence presented in Appendix A suggests that the Japanese firms use less capital per vehicle produced, so that the rate of profit measured as a return to capital would also be higher for the Japanese manufacturers. Thus, whether measured as a return on capital or as a margin of profit on sales, the Japanese producers earn higher profits on their U.S. sales than their U.S. counterparts.

Existing evidence suggests that in the late 1970s the Japanese achieved a noticeable edge in assembly quality ("fits and finishes"); since 1980, U.S. producers have made improvements in quality performance. Consumer ratings of vehicle condition at delivery and counts of defects per vehicle shipped in 1979, for example, show a significant import (i.e., Japanese) advantage; on a scale of 1-10, imports rated 7.9, while domestics averaged 6.4. When asked, "Would you buy the same make or model again?," 77.2 percent of domestic subcompact buyers answered yes; among import buyers the comparable percentage was 91.6.

Despite the popular image of Japanese superiority in advanced technology, explanation of the Japanese productivity advantage seems to be more a matter of differences in management--process systems, workforce management--than superior automation or faster work pace. Because of a production control system that emphasizes minimum inventory and elimination of downtime and a job structure that places responsibility for quality on workers, the Japanese operate processes at a high level of good output over extended periods of time. While several elements of the Japanese system are refinements of practices developed in the United States, certain critical aspects of their approach are accorded much less emphasis in U.S. practice. The policies and procedures connected with workforce management are a case in point.

The labor-management relationship established in the 1930s had its roots in the early years of the industry. The innovations in machinery and process design of the World War I era were accompanied by a system of workforce management characterized by highly structured rules and procedures. Planning and control of work were vested in staff groups far removed (organizationally) from the process. Workers were not involved in production beyond a narrow range of assigned tasks. The principal connection between the worker and the firm was the supervisor, and the relationship was essentially adversarial: supervisors were under pressure to meet production and cost targets, and that pressure for production at low cost was transmitted to the work force.

Unionization of the industry in the 1930s introduced a system of industrial jurisprudence into the workplace and changed the terms and conditions of employment in many ways. But the basic relationship between the worker and the firm remained adversarial in nature. Changes in the character of competition in the 1970s have highlighted weaknesses in that kind of relationship: it inspires no loyalty or commitment, and it fails to tap information and experience in the work force.

The last few years witnessed important changes in the employment relationship. Since the early 1970s, General Motors (GM) and the United Auto Workers (UAW) have worked to develop "quality of working life" programs; various approaches have been developed and extensively diffused in the organization. During the past year, "employee involvement" programs have been initiated at over half of Ford's facilities; Chrysler also has developed such efforts in connection with its quality-improvement efforts.

The kinds of changes under way are akin to a cultural revolution; where attitudes are deep seated, a true reformation is likely to require some period of time. Yet recent events suggest a good measure of adaptability in the collective bargaining relationship and thus reason for optimism.

TECHNOLOGY AND COMPETITION

A key issue separating the alternative interpretations of the industry's present and future condition is the role of technology in competition and the character of innovation. The "transient" and the "maturity" perspectives assume a stability in technology, that is, a relatively standardized technology that changes only incrementally and that is competitively neutral.

This description fits the condition of the industry prior to the initial OPEC shock of 1973, but there is some evidence that the role of technology in competition is shifting. Using the data before (1977) and after (1979) the Iranian oil shock, we find a substantive shift in the market's valuation of technology. Performance and technology characteristics associated with new designs (package efficiency, driving range, diesel engines, front-wheel drive) carried premiums in 1979, while the same characteristics were discounted in 1977.

In terms of market premiums the evidence implies that technology became more visible in the aftermath of the Iranian oil shock and a more important aspect of competition.

If technology becomes a more critical element of competition, innovation is likely to become more rapid and fundamental. Indeed, it appears that the development of product technology in the 1970s constitutes a sharp reversal of the pattern of technical change that dominated from 1900 to 1950.

The earlier era was dominated by standardization: first in engines, then the chassis, then components. In the 1970s, however, innovation spawned diversity in engine configuration, control systems, drive trains, and materials.

The pattern of technical development suggests that innovation is becoming less incremental in its impact on the production unit. Recent changes have not just refined existing ideas but have also introduced new concepts; downsizing, trans-axles, and new materials are examples. Future technologies carry the possibility of significant change in production facilities; advanced engine concepts, materials, and control systems require radically different equipment, skills, and organization.

Increased diversity and increasingly radical innovation leave a key assumption of the "transient" and "maturity" perspectives-- stability in technology--open to question. Because future development is uncertain, and because some systems (at least in small cars) have achieved dominance (front-wheel drive, four-cylinder engines), it is not possible to make precise and definitive statements about the course of technical change. If the incentive for innovation remains strong, however, it is likely that the market will see increased diversity of technology as new designs in engines, bodies, and other systems compete for market acceptance. If so, we may be at the beginning of a period of intense technology-based competition.

CONCLUSION:
THREE SCENARIOS AND THEIR IMPLICATIONS

This report identifies the historical context and the industry's current position in terms of product cost and quality and technology. Evidence on the three main lines of interpretation is presented, but the report draws no strong conclusions. Some evidence in favor of all three interpretations has been found, and there have been a number of assumptions made along the way. To further identify the implications of alternative patterns of development, the concluding chapter of this report presents three scenarios of the industry's future based on the three lines of interpretation.

The scenarios depict possible chains of events and the likely impact of those events on broad public policy options. While the scenarios are intended to offer a realistic assessment of the development of the industry under given assumptions, they are not based on an extensive analysis of business strategy. And although some very general views about public policy are indicated, an in-depth analysis of policy options was not carried out. The strategies of particular firms and detailed policy analysis are important areas for further work but were outside the scope of this report.

The three scenarios have quite different predictions for the future evolution of the industry.

Transient Economic Misfortune: The United States maintains a viable domestic industry, but the U.S. share of value-added declines; competition occurs much as before on the basis of styling, scale economies, and distribution.

Natural Consequence of Maturity: Local content of U.S. sales declines substantially; 65 percent of all cars sold in the United States are produced in foreign countries; U.S. firms survive but with substantial offshore production and only specialty vehicle production in the United States.

Fundamental Structural Change: Industry moves from full-line products and cost competition to more performance-oriented competition; the United States recoups market share with innovative vehicles, but the U.S. share of value-added declines because of losses in standard models.

Using two general categories of policy measures ("internal"-- deregulation, tax incentives; "external"--temporary policies to reduce imports), it is clear that predictions about the impact of policy depend on what scenario is assumed to pertain.

Internal policies have a major impact under the "transient" scenario, while both internal and external policies have a major impact under "restructuring." However, without permanent restrictions on trade, policy has no lasting impact under the "maturity" scenario; cost disadvantages in standard models are too large to be overcome through investment.

As in the case of public policy, implications for management's competitive and organizational agenda are somewhat different under the three scenarios. Under "maturity" the key to competition is the ability to manage a worldwide production and distribution system, with worldwide sourcing and technical innovation that extends and refines existing concepts. Under "restructuring" the essential tasks are improving quality and productivity in existing models and the development and introduction of radically new products and processes. These differences in competitive environment should not be glossed over, but it is also clear that both of these challenges require substantial changes in the way the business is managed. Some critical elements of that change are as follows:

* An emphasis on manufacturing as a major competitive factor.
* A more open agenda between management and labor.
* A move to engage the work force (all levels) in the competitive activities of the firm.
* An increased emphasis on the management of change; greater adaptability and openness to innovation, both organizational and technical.

For both public and private policy, prediction about what will be effective depends fundamentally on what is assumed about the industry's development. Both carry the potential for significant influence on the future of the industry. The future of the industry is by and large in the hands of its participants--the firms, the unions, the suppliers--but public policy has a critical supporting role to play, particularly in mitigating risks and facilitating necessary change during the period of transition.

1

The U.S. Auto Industry in Crisis

The U.S. automobile industry is in a crisis. Vigorous import competition, drastic shifts in consumer preferences, and anemic final sales combined to make 1980 and 1981 two of the most difficult years in the industry's history. The current picture is bleak: literally hundreds of thousands of people have lost their jobs; communities dependent on the industry have suffered devastating losses in employment and financial resources; all the domestic producers have suffered major financial losses; large facilities have permanently closed. Future prospects are uncertain. If the industry is to survive, the next five years will see wrenching changes in its productive and financial base as new product technologies are introduced; manufacturing plants are retooled; and new relations are established among management, labor, and government.

Even in a time of general economic malaise, trouble in the auto industry carries special weight. For more than half a century, the automobile--both as an artifact and as a business nexus--has played a significant role in the social and economic life of the nation. It uses 42.3 percent of all the oil consumed in America and accounts for roughly 15 percent of the average household budget. The factories that produce it and the businesses that service it employ a full 15 percent of the working population. When sales reach the low levels experienced in the last two years, the effects on employment and on communities where automobile production is important can be substantial, as the thousands of people on indefinite layoff and the record high unemployment rates in numerous midwestern cities clearly demonstrate.

Given the size and importance of the industry, past and present, it is little wonder that political leaders have long accorded it unusually close attention. Nor is it any surprise that concern for the future has made it among the most regulated of industries. But to view the automobile business as the most

"American" of our heavy industries, accurate as that view might be, is still to understate its place in our national life.

The industrial base that has grown up around automobile production possesses an immense strategic value of its own. Because of its size and technological sophistication, the manufacturing capacity of the industry can be turned, as it was during World War II, to the production of military equipment. And as recent events have shown, even the coming of the nuclear age has not diminished the critical reliance of the military on electromechanical equipment.

The strategic value of that industrial base is by no means limited to such applications. Ongoing process innovations, which enhance current products and make possible the creation of new ones, and competitive pressures for efficient production have made the industry a prime consumer--and a major stimulant--of technological advance. In recent years the auto companies have played a key role in the evolution of CAD/CAM (computer-aided design/computer-aided manufacturing), laser technology, new materials, industrial robots, and a host of other such developments. As Abernathy (1980) has argued, the existence of a set of customers demanding high performance in their cutting-edge technology and deeply committed to such innovations in their early stages are often of determinant importance in the development of new technology.[1]

It is, therefore, of no little consequence to the nation when the automobile industry finds itself in trouble. As in the past with good fortune, so with present problems, it has presaged change in other sectors. Defining that trouble accurately, pinpointing its causes, and prescribing appropriate remedies have within the recent past come to occupy a prominent place on the public agenda. Heated debate about the automobile industry is of course not new, but it has taken on a new urgency during the last two years.

In an important sense this escalation of argument is the direct result of events in the oil market during 1979. Although OPEC and rising oil prices have affected the industry since the oil embargo of 1973, the revolution in Iran marked a genuine turning point. OPEC seized the opportunity presented by substantial reductions in supply and strong upward pressures on spot market prices to double (and, in some cases, to more than double) the price of crude oil. As a result, gasoline prices in the United States rose sharply throughout 1979. Even so, there were widely publicized lines at U.S. gas pumps during the spring and summer.

In effect, these developments laid to rest any lingering hope that the power of OPEC was on the wane or that oil prices might fall significantly in the future. With changed expectations about the future course of the price of gasoline and heightened concern about interruptions of supply, American consumers abruptly demonstrated a shift in market preference toward smaller, more fuel efficient cars--sometimes domestic, if available; but imported, if not.

Some have argued that the unexpected surge in gasoline prices coupled with a shift in consumer preference away from intermediate or larger cars is primarily responsible for the current difficulties in the industry. In testimony before the Subcommittee on Trade of the House Ways and Means Committee, Abraham Katz, Assistant Secretary of Commerce for Trade, made the following observation:

> Early in 1979 . . . a sudden disruption in OPEC oil shipments and large OPEC price increases led quickly to sharp increases in the price of gasoline and to renewed gas station lines. . . . Consumers reacted by shifting toward small, fuel-efficient cars. Small car sales jumped to a 57 percent share of the market in 1979. U.S. small car production ran virtually at capacity, but was unable to keep up with demand. With an inadequate supply of domestic small cars, many consumers turned to imports, the traditional source of small, fuel-efficient cars. Their present success in the United States is a case of being in the right place at the right time with the right product.[2]

It is also apparent that a good part of the U.S. auto industry's plight reflects the overall state of the economy. Automobile sales are sensitive to changes in interest rates and the growth of real income. The decline in real income in recent years, high interest rates, and generally sluggish economic activity have reduced demand for automobiles to very low levels. Coming at a time when major changes in product mix and new capital investments are required, the recession has made adjustments much more difficult. Thus, while the gasoline price shock of 1979 and shifts in consumer preferences may have affected the relative demand for domestic production, the low overall level of demand must be weighed as a major factor.

It would be unwise to assume that the only problem is a lack of market growth. Moreover, as far as competition with imports is concerned, more is involved than simply the size of American cars or their fuel efficiency. Perceived differences in product quality between domestic and imported cars also are at work. Some analysts have suggested that a comprehensive statement of the industry's problems must start with the recognition that, as in American industry generally, lack of investment, a faltering work ethic, excessive regulation, and the declining growth in productivity are all responsible in one degree or another for deteriorating product quality.

The firms, the UAW, and the government have all been the subject of criticism in the public debate over the industry's condition. Failure to exploit long available technology, generous collective bargaining agreements, inattention to market development, a confusing welter of regulation, high absenteeism, artificially low gasoline prices, among other things, have all been cited in one place or another as causes of the current situation. Whatever the mix of truth and error in all this finger pointing, one thing is reasonably clear: Detroit's troubles are not just the result of any single discrete, isolated cause, such as an inappropriate product mix. To understand the difficulties accurately, we must focus our attention on the interplay of causes within the whole complex productive federation of the industry. We must seek to understand the roles played by each of the participants in that federation and, more than that, the ways in which the decisions of one influence and affect all the others.

Accordingly, we intend to address ourselves in this report to the general competitive status of the U.S. automobile industry. Though we recognize the effects of oil prices, regulation, and widespread economic stagnation on the fortunes of the industry, we also feel that much of its current plight is the result of factors internal to the industry and its productive confederation. To say this is not to argue that such things as the doubling of gasoline prices in one year are of little moment. It is, instead, to argue that the situation is an exceedingly complicated one--one that cannot be ameliorated simply by a realignment of the standard product line of U.S. manufacturers. The capacity to innovate successfully, in technology and in organization, is also necessary if those companies are to be truly competitive on an international basis.

We cannot, however, undertake this report as if alternative interpretations of the industry's present and future condition were not already available. At the risk of some over-simplification, we have organized those lines of interpretation into three distinct groups and have structured the report so as to sort out the evidence that bears on them and on their underlying assumptions. The first of these categories of interpretation we have labeled "transient economic misfortune"; the second, "natural consequences of maturity"; and the third, "fundamental structural change." A brief word about each is in order.

The first view, long popular with officials of the Carter administration, is that the current crisis in the automobile industry, though a serious misfortune, is nonetheless temporary. Since consumers shifted to imports only when the domestic producers were unable to supply enough small, fuel-efficient cars and since the expansion plans of the domestic producers are known in advance, the end of the crisis can be predicted quite accurately. If American manufacturers can, as planned, turn out between 6.5 and 7 million small cars per year by 1985, American consumers will happily return to the fold--or so runs the argument of "transient economic misfortune." Missing, of course, is any convincing reason to believe that the domestic producers know exactly what to produce, that what they produce will be competitive, or that their competitiveness--even if achieved--will persist into the future.

A different interpretation is offered by those who see in the current problems of the auto industry the "natural consequences of maturity." This view, based on theories of the product life cycle in international trade, treats the development of such products as the automobile, computers, or television sets as a predictable sequence of stages from an uncertain technological "infancy" to a highly standardized technological "maturity." As production requirements change over the course of the life cycle, countries that enjoyed an advantage in the early stage of evolution will lose it at a later stage. Indeed, one of the main predictions of life-cycle theory is that the location of production will shift over time as the product matures and its technology diffuses.[3]

Considered in these terms, the automobile industry is rapidly approaching a mature state, a state in which both product and process technology are stable and well known. As a result, competitive advantage depends less on significant advances in product development than on relative costs of production and of production factors. By rights, the locus of production ought to shift from those countries where factor prices (labor, capital, materials) are high to those where they are low. This scenario has already been played out in other industries, such as textiles, motorcycles, TV receivers, and radios. Why not, then, with automobiles?

In fact, from the standpoint of this "maturity" view, the only thing remarkable about the problems facing American auto manufacturers is the timing of the surge of imports. The precipitate rise in imports may have been unexpected, but the long-run tendency for them to displace domestically produced cars was entirely predictable. Imperatives of cost may still leave domestic producers with specialty-market niches, but the logic of "maturity" argues that in time the bulk of demand will inevitably be met from low-cost sources.

Both these lines of interpretation envision a growing stability in product technology but differ in their assessments of the domestic producers' ability to compete. Proponents of the first line of interpretation believe that domestic cost disadvantages can be overcome by appropriate capital investments; proponents of the second believe those disadvantages to be inherent and permanent. Proponents of the third interpretation, the view we call "fundamental structural change," challenge the assumptions on which the first two rest, for they deny the fact of stability in product (or process) technology.

To this third group the events of the late 1970s marked the beginning of a new era, an era in which the incentives and the rewards for technological innovation increased dramatically. In this view, radically different power plants, drive trains, body structures, and control systems loom on the horizon, promising advantage to those companies capable of technological leadership. The old stable order has been overthrown by events, and a new future of great technological diversity awaits the talented and the bold.

If competition in the automobile industry between 1945 and 1978 occurred within well-defined technological limits and was dominated by marketing, styling considerations, and economies of scale, the proponents of the fundamental change view argue that competition in the 1980s and beyond will, once again, be heavily influenced by technological innovation. In this respect the industry will become much more as it was in its early years when product technology was changing rapidly and significant competitive advantage accrued to those who introduced major functional innovations. If this last view is correct, there will be a "greening" of the automobile industry, a period of striking industrial "de-maturity," in which the technology is diverse, uncertain, and changing.

It is to the examination of the relative merits and implications of these three lines of interpretation that we now address ourselves. Chapter 2 of our report provides something of a primer on the industry. In it we sketch out the basic facts of the market, the production process, and the companies.

Following the industry primer, Chapters 3, 4, and 5 examine three historical trends that in retrospect have been of critical importance in the industry's development up to the beginning of the present crisis in early 1979. These include the covergence of technology (Chapter 3), the internationalization of products and markets (Chapter 4), and the growth of goverment regulation (Chapter 5). With the historical developments as background, we shift to an analysis of the current situation, seeking to understand developments in the market, the competitiveness of domestic products, and the role of technical advance. Chapter 6 focuses on the comparative cost and quality of U.S. products and Chapter 7 on the implications for the management of people. The role of technology in competition is examined in Chapter 8, and Chapter 9 considers the nature of recent technical innovations. Finally,

Chapter 10 develops alternative scenarios of the industry's future and discusses their implications for public and private policies.

NOTES

1. See Abernathy (1980) for a full discussion of these issues.
2. Katz (1980).
3. Wells (1980) makes this argument clearly.

2

An Industry Primer

The automobile has become an integral part of everyday life for millions of people. What was once looked upon with wonder and awe has become commonplace, a durable consumption good taking its place alongside countless other gadgets and machines that inhabit modern garages and households. Thousands still flood to auto shows to see new and exotic hardware, but the technology in the basic run-of-the-mill automobile is taken for granted.

Yet the car is among the most sophisticated, complex consumer products ever devised. Early gas-powered vehicles were little more than a modified carriage with a crude (by modern standards) engine and chain drive; "horseless carriage" was a quite accurate description. Many years of refinement and development have resulted in an engineering- and technology-intensive product. Major technical systems include the engine with its advanced mechanics and materials, fuel delivery with sophisticated carburetion or fuel injection, automatic transmission and drive train, power-assisted steering and brakes, and complex electronic controls.

The technical complexity of the car is masked by the simplicity of its operation. From the standpoint of the driver, all that is required is a turn of the key, selection of forward or reverse, and pressure on the gas pedal. Beneath the sheet metal, behind the gear selector, however, the technical systems must function at high levels of performance, on demand, under extreme conditions, over and over again.

From the beginning of the industry, reliability under pressure and simplicity of operation were important parts of the motivation for increasing complexity and sophistication of the car's technical systems. Reliability, simplicity, and low cost were essential to the development of a true mass market oriented toward basic transportation. In this sense, design changes in the first 20 years of the industry were determined by market demands, and competitive success depended on significant advance in

17

function and performance. At this stage, much of the new-car demand was "first time" purchase, with replacement demand playing a relatively small role.

The rapid and widespread acceptance of the automobile is evidence of the success of engineering and technical developments. As the product matured and basic technologies were refined, the design of basic systems and components was stabilized. The previously dominant need for basic transportation gave way to a more varied, more sophisticated set of demands and consequent segmentation of the market. Vehicles were developed to meet particular functions (e.g., station wagons, sports cars, family sedans). Moreover, within a given function, the use of optional equipment created wide divergence in the potential cost and performance characteristics of similar models.

Table 2.1 presents a four-segment characterization of the automobile market of the 1970s. The segments range from

TABLE 2.1   Market Segments (by model) and Buyer Priorities in the 1970s

	Market Segments			
	Economy Cars	Sporting/ Personal Cars	Family Cars	Prestige/ Luxury Cars
	Chevette	BMW 320i	Fairmont	Audi 5000
	Civic	Camaro	Impala	Jaguar
	Corolla	Celica	LeBaron	Lincoln
	Omni	Cutlass Coupe	LTD	Mercedes-Benz
	Pinto	Grand Prix	Regal	Seville
	Rabbit	Mustang		
		Thunderbird		
Priorities				
Price	x	x	x	
Fuel economy	x			
Interior room			x	
Comfort			x	x
Reliability	x		x	x
Acceleration		x		
Handling		x		
Styling		x		x
Interior trim		x		x
Workmanship				x

SOURCE: Adapted from Arthur D. Little, Inc., *The Changing World Automotive Industry Through 2000* (1980, pp. 24-25).

TABLE 2.2   Structure of New Car Sales

Year	Size Class				
	Subcompact[a]	Compact	Intermediate	Standard	Luxury
1967	9.3	15.7	23.6	47.9	3.1
1972	22.7	15.4	21.7	36.1	3.4
1973	24.9	17.7	23.0	30.0	3.6
1974	28.4	20.0	24.2	22.6	3.7
1975	32.4	20.3	24.1	17.9	4.0
1976	26.1	23.5	27.3	19.4	3.7
1977	27.1	21.2	26.9	19.4	4.6
1978	26.4	21.6	26.8	18.4	5.5
1979	34.0	20.0	24.2	15.3	5.5
1980[b]	42.0	20.2	20.6	12.5	4.7

[a] Includes imports.
[b] January and February.

SOURCE: *Ward's Automotive Year Book,* Annual Reports, Detroit, Michigan.

economy car to luxury/prestige car and are defined primarily in terms of consumer preferences and principal use. The table also contains an assessment of the priorities governing purchase decisions in each segment. Both the segments and the priorities reflect conditions in the mid-1970s before the oil crisis of 1979. Given changes in the relative price of fuel and in household types over the next several years, buyer priorities and segmentation are likely to change. Patterns of change expected for the future are somewhat evident in the shifting pattern of demands by size class over the last 10-12 years.

Although not a perfect measure of the diversity of demands in the postwar era, vehicle size has been an important competitive dimension. Larger cars have been associated with luxury, elegance, and prestige, and many important product innovations were first developed for larger cars and then diffused to the rest of the product line. The dominance of the large car reached its zenith in the late 1960s. With the onset of higher operating costs the structure of demand in terms of vehicle size has shifted dramatically.

Table 2.2 presents data on new car sales by size class for selected years since 1967. The data document the sizeable shift in the structure of demand that occurred in the 1970s. In 1967 over 70 percent of new car sales were in the intermediate and standard categories, while the subcompact group, which includes imports, accounted for 9.3 percent. By early 1980 the subcompact group dominated the market, while standard models held fast at 12.5 percent of the market.

The shift to smaller cars has been well publicized, but the timing of the change and the patterns of adjustment within the large-car ranks (intermediate, standard, luxury) deserve emphasis. It is clear from Table 2.2 that the largest change in the share of standard models occurred in 1974, but the downward trend was evident long before OPEC quadrupled the price of crude oil. From 1967 to 1972, for example, the large-car share fell from 71.5 to 57.8 percent, with most of the decline coming from the standard group. At the same time the share of subcompacts more than doubled to reach 22 percent. The shifts evident in the pre-1973 data set the pattern for the rest of the decade: a dramatic decline in the share of standard-size vehicles, a rise in subcompacts (including imports), and only modest changes in the share of other categories.

THE MANUFACTURING PROCESS

Changes in the structure of demand in the 1970s have had a profound impact on manufacturing facilities and processes. Transforming equipment, plants, and organization geared to the production of large road cruisers into a system for the design of much smaller and more efficient cars cannot be accomplished overnight. The expense and difficulty reflects the complexity, scale, and integration of the automobile production process. Indeed, the complexity and sophistication of the car itself pale in comparison to the organization and technology used to design, produce, and deliver the finished product to the market. Manufacturing involves the production or purchase of over 10,000 parts, combining parts into components and systems, and the coordination of all this activity so that the right systems and components can be assembled to produce an automobile. The basic structure of production in the industry is shown in Figure 2.1. The figure includes activities in the chain of supply from raw materials to final assembly.

Perhaps the dominant characteristic of the automobile production process is the importance of economies of scale. Over the last 70 years the production process has become increasingly mechanized, automated, and capital intensive. Indeed, the classic illustration of automation in U.S. industry is the modern automobile engine plant.

The modern engine plant can be seen as the outcome of an evolutionary progression from the general-purpose job-shop environment that characterized early engine manufacture. The highly specialized, capital-intensive process in today's plants

PROCESS SEGMENT

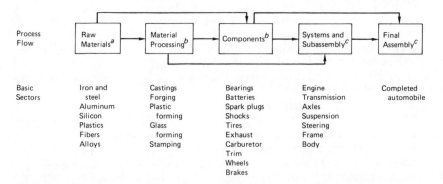

FIGURE 2.1 The structure of production in the automobile industry. (Adapted from Byron, 1980.)
[a]Predominantly suppliers.
[b]Mixed OEM/suppliers.
[c]Predominantly OEM.

reflects a strategic orientation toward low-cost production of a standardized product. Choices about equipment and process have created a setup in which high volumes are essential to low cost. Existing estimates of the minimum efficient scale in engine production range from 350,000 to 500,000 units per year, depending on the particular technology employed.

Engine plants are more automated than many of the processes in automobile manufacture, but the dependence of low cost on high volume is characteristic of all of them. In general, manufacturing policy has been oriented toward increased standardization and specialization in manufacturing operations and consequent reliance on high volume. At the same time, employment in jobs not directly related to production has increased substantially, possibly reflecting the growing complexity of coordination problems and changing regulatory requirements. In addition, the fixed component of engineering and research and development costs has grown under regulatory pressure and the need for new design initiatives. The net effect of these developments has been to enhance the importance of scale in the determination of profitability and competitive advantage.

The Plant Network: An Illustration

The basic structure of the overall manufacturing process, and in particular the plant network, can be illustrated by considering the impact of changes in the marketplace as demand has shifted to smaller, more efficient cars.[1] The shift in demand has been met by "downsizing," by changes in basic components (e.g., shift from rear-wheel to front-wheel drive) and by material substitution. Though each change may have a direct impact on only part of the car, or a part of the manufacturing process, the various types of facilities are so tightly linked that even a small change can have major ramifications.

The key facilities in the manufacturing process and their linkages are illustrated in Figure 2.2. Though highly simplified, the diagram captures the basic relationships among the manufacture of materials (e.g., steel, aluminum, plastic), components (e.g., steering gears, brakes), systems (e.g., engines, transmission), and final assembly. The automobile companies [hereafter, OEMs (original equipment manufacturers)] do final assembly in their own facilities, and they generally produce major systems (engines, transmission) in-house; most materials and many components (e.g., brakes, steering assemblies, valves) are purchased from suppliers. The extent of integration varies significantly by company and even by model. It is not uncommon for OEMs to produce part of their need for a component in-house, while maintaining additional sources outside.

In Figure 2.2 the impact of "downsizing" is indicated by the cross hatches on various facilities. "Downsizing" involves all new body sheet metal (retool stamping plants), a shift to V-6 engines (change engine lines), smaller and lighter components (retooling at several plants--axles, suspension, brakes, and so forth), and new frames. The changes culminate in the final assembly plant, which requires some new fixtures and tooling. It is not hard to see why such changes require several hundred million dollars and take a few years to accomplish. Yet "downsizing" has a relatively modest impact on the manufacturing process in comparison to the redesign of basic components. The impact of moving from rear- to front-wheel drive and the introduction of unit body construction are illustrated in Figure 2.3. These changes in drive train and frame are accompanied by changes in several major components, as follows:

• Elimination of the rear axle and addition of a new rear suspension.
• Replacement of the standard V-8 engine with V-6 and IL-4 engines.
• Elimination of the drive shaft and transmission and replacement with trans-axle and twin front-wheel-drive shafts.
• Addition of new suspension (McPherson struts) and steering (rack and pinion).

23

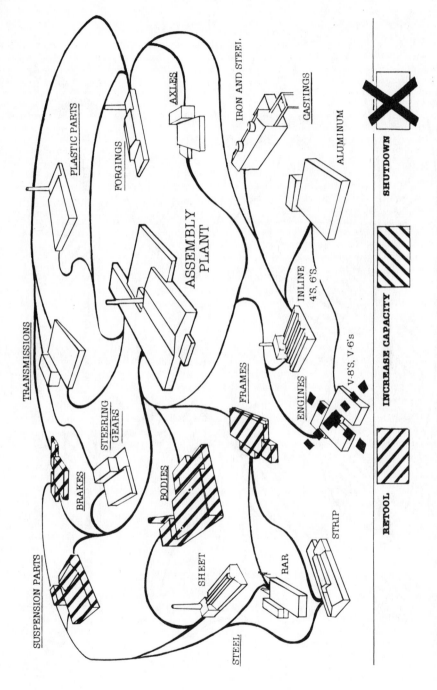

FIGURE 2.2 The impact of downsizing on production facilities. (Adapted from Byron, 1980.)

24

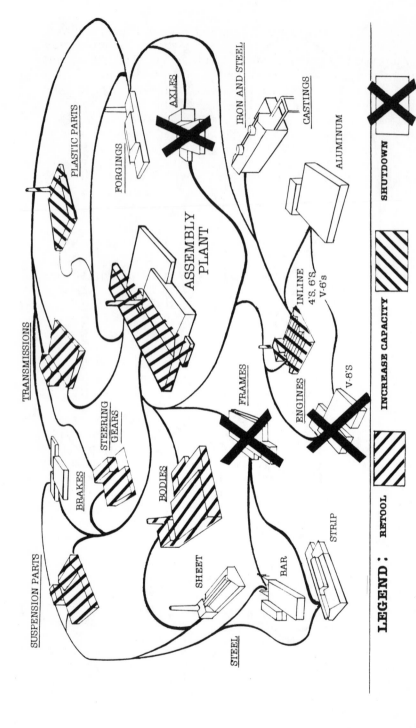

FIGURE 2.3 The impact of basic component redesign on production facilities. (Adapted from Byron, 1980.)

The impact on facilities is widespread: engine, frame, and a plants are shut down, and major retooling occurs in the productio of transmission bodies and other components. The significant capital adjustments, however, involve only modest use of new materials and do not require major changes in manufacturing processes. Future changes in design and in the use of new materials could have even more far-reaching effects. We have already seen the introduction of electronics, which adds another type of process and facility to the overall system. Increased use of plastics, composites, and lightweight metals could make older facilities and processes obsolete and require new manufacturing capabilities.

COMPANIES AND STRATEGIES, 1908-1973

The scale and complexity of the auto production process and the emphasis on high volumes is not a recent event. Although the very early days of the industry were characterized by competition among small technological entrepreneurs, the emergence of the Model T in 1908 and the subsequent development of a mass market for automobiles gave strong impetus to the emergence of large-scale enterprise as the dominant form of organization.[2] By 1923 Ford and General Motors (GM) held 71 percent of the market, with Ford's share amounting to 50.4 percent. Ford lost the leadership position to GM in the late 1920s, and GM has retained a dominant market share to the present time.

From 1925 to 1970, competition in the auto industry was essentially competition among a few giant domestic firms. While several so-called independents operated at the margin of the industry until the early 1950s, the bulk of sales was satisfied by the Big 3: GM, Ford, and Chrysler. The nature of competition in this period was strongly influenced by the strategy developed by GM in the 1920s and 1930s. In terms of pricing, product and process technology, and distribution, the Big 3 developed broadly similar approaches, although both Ford and Chrysler fashioned distinctive features.

Stated quite broadly, the history of competition in the auto industry up to the oil embargo of the 1970s was marked by two distinctive periods. Table 2.3 presents a brief characterization. In the first period, from 1908 to 1948, major innovative changes in the product played a significant role in the jockeying for profits and share. The second period was marked by relative stability in product technology and increased emphasis on competitive pricing and styling. In the postwar era, competition occurred primarily on the basis of economies of scale, styling, and the dealer network. These broad evolutionary changes are reflected in the changing strategic orientation of the major firms.

ging Mix of Competitive Factors

	Stage of Development	
	Early (1905-1948)	Late (1949-1973)
...ng	*Secondary Factor* Product performance dominates price comparisons; initial buyers value performance over price.	*Primary Factor* Standardization leads to acceptable levels of performance; price becomes significant factor in purchase decision.
Model change (innovation in technology)	*Primary Factor* Significant improvement in product occurs rapidly; new models have major impact on market share.	*Secondary Factor* Technology is refined and standardized; new models offer styling changes.
Channels of distribution (dealerships)	*Primary Factor* Personal contact and dealer reputation are key to acceptance of new product.	*Primary Factor* Availability, cost, and quality of service are important to mature product.

SOURCE: Adapted from Abernathy (1978, Table 2.5, p. 41).

Strategy at Ford

The innovations in product and process that carried the Ford Motor Company to a dominant market position between 1908 and 1927 were motivated by a broad strategic plan. The essential outline of Henry Ford's strategy is suggested by an advertisement he placed more than two years before the Model T was introduced:

> [The] idea is to build a high grade, practical automobile that can be maintained as near $450 as it is possible to make it, thus raising the automobile out of the list of luxuries and bringing it to the point where the average American citizen may own and enjoy his own automobile-- the question is not "how much can we get for the car?" but "how low can we sell it and make a small margin on each one?" [3]

The design of the Model T was followed by Ford's innovations in process methods and decentralized assembly plants, with mass production and distribution to provide control of the markets in an era of slow communications. The success of the strategy was evident in dramatic price reductions and in expansion of the market from 1908 to 1926; by 1923 Ford had 50.4 percent of a market that had grown to 3.6 million units.

By the early 1920s the Model T competed in a market far different from that of 1908. Its design had been improved upon, and the lack of variety had given GM an opportunity to differentiate and segment the market. Even though Ford added a starter and a closed steel body in the mid-1920s, there was no change in basic design. To retain market share, Ford dropped the price to $290, but GM still gained market share rapidly. Ford closed down completely in 1926 for nine months to design and change over to a new model.

Ford introduced a new product in 1927 (the Model A), but the strategy was unchanged. Although Ford briefly regained its prior market share, the old competitive approach of low price, standardized design, and mass production did not work for long. After three years, Ford's market share dropped below 25 percent. Product standardization was abandoned in 1932 with the introduction of the V-8 engine.

Alfred Sloan of GM criticized Ford's strategy as follows:

Mr. Ford had unusual vision, imagination and foresight-- [his] basic conception of one car in one utility model at an even lower price was what the market, especially the farm market, needed at the time. . . . [His] concept of the American market did not adequately fit the realities after 1923. Mr. Ford failed to realize that it was not necessary for new cars to meet the need for basic transportation. . . . Used cars at much lower prices dropped down to fill the demand. . . . The old master has failed to master change.[4]

Ford's strategy was brilliant but rigid. A market need was identified; the product and the manufacturing, marketing, and distribution facilities to meet the need were developed and implemented. But Ford's strategy recognized neither the dynamics of market development nor the counteractions of competitors.

Under new management after World War II, Ford rapidly adopted a new strategy.[5] Independent divisions, each having its own product lines and production facilities, were envisioned. Separate engine and assembly plants for Lincoln-Mercury and Ford divisions were introduced, but the market failure of the

Edsel thwarted the planned development of three separate car divisions. After 1960 all North American production facilities were consolidated under a centralized functional organization; that is, many of the same production and engineering functions served all product lines.

In describing competitive policies, Lawrence J. White concludes that Ford has been a follower in styling but a leader in seeking out market niches.[6] New models like the Mustang, Maverick, Pinto, and a combination car and truck called the Ranchero seem to confirm this characterization. Despite these successes, Ford has not been able to excel in head-on competition with GM across the full product line.

Alfred Sloan and GM's Strategy

GM's competitive policies evolved out of experience with both success and failure in the contest with Ford. The basic approach has been summarized by Alfred Sloan:

> In 1921 . . . no conceivable amount of money, short of the United States Treasury could have sustained the losses required to take volume away from [Ford] at their own game. The strategy we devised was to take a bite from the top of his position--and in this way build up Chevrolet volume on a profitable base.
>
> Nevertheless--the K Model Chevrolet--was still far from the Ford Model T in price for the gravitational pull we hoped to exert in Mr. Ford's area of the market. It was our intention to continue adding improvements and over a period of time to move down in price on the Model K as our position justified it.
>
> We first said that the corporation should produce a line of cars in each price area, from the lowest to one for the strictly high grade quality-production car. . . . We proposed in general that General Motors should place its cars at the top of each price range and make them of such quality that they would attract sales from below that price. . . . This amounted to quality competition against cars above a given price tag and price competition against cars about that price tag. . . . The policy we said was valid if our cars were at least equal in design to the best of our competitor's grade, so that it was not necessary to lead in design or run the risk of untried experiment.
>
> The same idea held for production--it was not essential that for any particular car production be more efficient than that of its best competitor--coordinated

operation of our plants would result in great efficiency--
the same could be said for engineering and other
functions.[7]

Thus, there were three essential elements in GM's strategy: (1)
Product design was conceived as a dynamic process that would
lead to an ultimate target through incremental change. Design
was not a once-and-for-all optimization as it had been with Ford.
This process later became the annual model-change policy of GM.
(2) Market needs would be met through the product-line policy
rather than independent designs. (3) Radical product innovations
were to be avoided. As Sloan said, it was "not necessary to . . .
run the risk of untried experiment."

The broad competitive strategy that GM hammered out in
specific decisions was to prove unbeatable. The company gained a
dominant position in the U.S. market in the 1920s and has held it
to the present. Little change in the essentials of GM's strategy
was apparent during the period 1923-1973. Increased centraliza-
tion among operating divisions, less difference in technological
characteristics of various cars in the product line, and greater
sharing of common components tended to make the different car
lines more like a single product. In general terms, however, the
strategy seems to have remained intact.

Chrysler and Product Engineering

The Chrysler Corporation seized a foothold in the market when
Ford faltered in the Model T program and shut down for nine
months.[8] By 1929 Chrysler offered four basic car lines: Chrysler,
DeSoto, Dodge, and Plymouth. Unlike GM, production for all
product lines was centralized, and Chrysler apparently did not
integrate vertically backward as extensively as either GM or
Ford. Because Chrysler produced fewer of its own components, it
was less constrained in adopting advanced innovative components.
Thus, Chrysler could seek competitive advantages through flexibility in product engineering and in styling. Chrysler pioneered in
high-compression engines in 1925; in frame designs permitting a
low center of gravity in the 1930s; and in the experimental introduction of disc brakes in 1949, power steering in 1951, and the
alternator in 1960.

This strategy of design flexibility and shallow vertical integration proved very successful in the prewar period, when the rate of
technological change in the product was rapid. As product designs
stablized after the war, however, other factors like the strength
of dealerships and economies of scale became more important.
Chrysler's market share followed a downward trend after World

War II. Chrysler did develop strength in some segments of the market (vans, compacts) but was generally a follower in product development after the war. Cost control was difficult during times of inflation, when cost increases could not be passed on to the consumer.

This aspect was particularly troublesome after 1970. Inflation, government price controls, and the consumer's loss of real purchasing power have squeezed margins and capital at the very time when resources have been needed to develop and introduce smaller, more efficient cars. Chrysler's product image has not been well defined, and it has suffered a loss of customer loyalty and sales potential. Its current financial difficulties raise serious questions about long-term viability as a full-line producer. A competitive strategy emphasizing flexibility in product design was well suited to prewar conditions. As with Ford's early policies, however, it would seem that the development of the industry changed the necessary conditions for success.

The Imports

Imports have played a major role in the compact and subcompact segments of the U.S. market since the late 1950s. Foreign producers, notably Mercedes, BMW, and Triumph, have been important in specialty and luxury cars. The distinguishing feature of import strategies has been their emphasis on uniqueness in selected market niches. Whether in terms of size, performance, or quality, foreign firms have sought an advantage by creating products that were different from the standard or traditional domestic products. Furthermore, the more successful firms have built strong sales and service networks.

The clearest example of the importance of the dealer network in entering the U.S. market is the case of Volkswagen (VW). Firmly established before sales were made, the VW system of dealers became a distinctive competitive factor, particularly in comparison with other European manufacturers. VW's strategy of "service first" allowed the company to maintain a strong market presence through the 1960s. When relative costs of production shifted in the late 1960s and early 1970s, VW established a production facility in Pennsylvania.

The lessons of the VW experience have not been lost on the Japanese or other Europeans. The major Japanese firms have payed close attention to the development of a dealer network. The Renault-American Motors Corporation (AMC) relationship is motivated in part by Renault's desire for an established dealer system. Furthermore, production of Renault designs in AMC facilities is likely within the next few years. Other foreign manufacturers, notably Honda and Nissan, plan to open car and light-truck production facilities in the United States.

Firm Performance in the 1970s: Response to Crisis

Historically, the auto industry as a whole has earned returns above the average for manufacturing, both in terms of returns on sales and stockholders' equity. At the same time, however, those returns have shown much greater than average cyclical variability. The decade of the 1970s witnessed a trend toward erosion of the profitability of the domestic producers and marked cyclical swings in the recessions of 1970, 1974-1975, and most recently in the 1979-1980 period. The downward trend in profitability may reflect declines in real income, rapid shifts in relative prices, an inappropriate product mix, and effects of increased price competition from imported products. The importance of price competition is evident in Table 2.4, which presents data on Ford's list prices expressed in constant 1958 dollars and cumulative units of production. The long decline in the real price of the Model T, from 1908 to 1926, is indicative of Ford's "experience curve" strategy.[9] With the transformation of the market in the late 1920s and GM's leadership in building larger and more luxurious cars, the real price rose for over 30 years. Since 1960 two dips have occurred, both associated with import competition. It seems clear from these data that part of the weakening financial performance of the domestic producers can be traced to declining real prices, caused in part by intensive competition.

The oil price explosions in 1973 and again in 1979 played a key role in setting the economic context of industry performance. The oil crises affected the major firms very differently. Table 2.5 summarizes the basic competitive positions and market performance of the major domestic firms and indicates some of the actions taken in the aftermath of the twin oil shocks. Except for GM, which has gradually increased its share in the last few years, the domestic producers have lost significant market shares. The loss of markets is a reflection of rapid market shifts and lags in response. Ford, Chrysler, and AMC have lagged behind GM in introducing strategic changes in vehicle size or new products. In general, GM has adopted a more aggressive posture, an approach consistent with traditional market leadership and greater financial resources.

Financial performance has deteriorated for all firms except AMC, for which 1979 was an exceptionally good year. All of the Big 3 have experienced declining margins, with Chrysler suffering

32

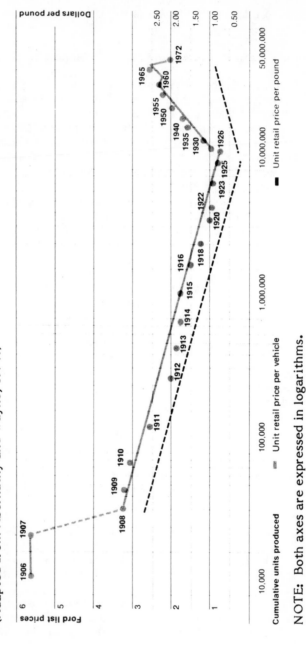

TABLE 2.4 List Prices and Cumulative Volume at Ford Motor Company, 1908–1972 (Adapted from Abernathy and Wayne, 1974.)

NOTE: Both axes are expressed in logarithms.

TABLE 2.5   Competitive Status and Firm Performance in the U.S. Auto Industry

Company	Dominant Strategic Orientation	Percentage of Market Share		Percentage of Return on Sales		Debt Ratio[a]		Response to Oil Crisis
		1970-1978	1979	1970-1978	1979	1970-1978	1979	
General Motors	A car for "every purse and purpose"; full-line producer.	44.6	46.5	5.6	4.4	6.4	4.0	Downsize top of the line; dieselization; new products (X body, J body); announced aggressive technology development.
Ford	Full-line producer; seek out specialty/unexploited niches.	23.8	20.3	3.4	2.2	15.5	11.0	Lag of 1-2 years in downsizing; product development (Escort).
Chrysler	Full-line producer; advanced product engineering; strength in vans/compacts.	12.9	9.0	0.5	-9.1	27.5	34.0	Lag in downsizing of large cars; product development (Omni, K cars).
AMC	Specialty small-car producer (jeeps, small cars).	2.8	1.5	0.5	2.7	20.4	11.0	Lag in product development and weight reduction; developed relationship with Renault.

[a] Long-term debt as a percentage of total capitalization.

SOURCE: Annual reports; *Moody's Industrial Manual* (cited in Arthur D. Little, Inc., *The Changing World Automotive Industry Through 2000*, 1980).

sizeable losses at the end of the decade. The addition of 1980 data would show negative returns for all producers, a situation that became evident in late 1979.

In retrospect, most of the 1970s was a time of transition for the U.S. auto industry. The once profitable, vigorous auto firms have experienced major financial and economic jolts; adjustments to the energy shocks and shifts in consumer tastes have not been smooth or easy. As we have intimated in this chapter, the nature of the U.S. firms' response to the crisis and the likely character of the industry's evolution in the next decade have been and will be strongly influenced by the specific pattern of industry development up to 1973. Several facets of that development might be examined fruitfully; we have chosen to emphasize three: the convergence of product and process technology, the international-ization of markets, and the growth of government regulation. Each is examined in subsequent chapters.

NOTES

1. The figures underlying this discussion are taken from Byron (1980).

2. This has been well documented; see, for example, Chandler (1964) and Abernathy (1978).

3. The original source is Nevins and Hill (1954), p. 282; also cited in Abernathy (1978), p. 33.

4. Sloan (1972), pp. 4, 186, 187.

5. For a more extended review of Ford's strategy, see Abernathy (1978), pp. 30-33, and Leone et al. (1981).

6. White, L. (1971), p. 207.

7. Sloan (1972), pp. 71-73.

8. For a discussion of Chrysler's strategy, see Abernathy (1978), pp. 36-38, and Leone et al. (1980).

9. See Abernathy and Wayne (1974) for an extensive analysis of Ford's "experience curve" strategy.

3
The Evolution of Technology:
From Radical to Incremental Innovation

Anyone trying to buy a car in 1905 was confronted with a bewildering array of products and technologies. There were cars powered by steam, electricity, or gasoline; cars with three or four wheels; cars with open-air cabs or closed carriages. These differences were not merely cosmetic. Structural features, mechanical principles, and performance characteristics varied widely from car to car.

Seventy-four years later, before the oil crisis of 1973, that technological diversity had all but disappeared. To be sure, the cars of the early 1970s displayed an immense, if superficial, variation in styling and model choice. But the underlying technology--the fundamental characteristics of structure and mechanical system--had become standardized. So, too, had the processes of production.

This evolution of the automobile industry from a state of technological diversity to one of standardization--and, for that matter, from a state of rapid and at times radical change to one of incremental innovation--is neither a random event nor an event peculiar to the automobile industry. The history of many industries and of many individual products shows the same development toward mature standardization from an earlier, more fluid condition.

The technological maturation of the auto industry, however, appears to be closely related to the nature of competition. In this chapter we contrast U.S. development with the quite different pattern in Europe. The evidence confirms the importance of competition and consumer tastes and suggests that government policy may affect the character of technological advance.

INFANCY TO MATURITY:
A PARADIGM OF TECHNOLOGICAL EVOLUTION

In general terms the evolution of a given product line and its associated production process(es) can be meaningfully described by

TABLE 3.1 Patterns of Product/Process Evolution

Dimensions of Performance	Stage of Development		
	Fluid	Transition	Mature
Characteristics of Production			
Process	Flexible and inefficient; major changes easily accommodated.	Becoming more rigid, with changes occurring in major steps.	Efficient, capital intensive, and rigid; cost of change is high.
Equipment	General-purpose, requiring highly skilled labor.	Some subprocesses automated, creating "islands of automation."	Special purpose, mostly automatic with labor tasks mainly monitoring and control.
Products/Markets			
Product line	Diverse, often including custom designs.	Includes at least one product design stable enough to have significant production volume.	Mostly undifferentiated standard products.
Competitive emphasis	Functional product performance.	Product variation.	Cost reduction.
Innovation			
Stimulus	Information on users' needs and users' technical inputs.	Opportunities created by expanding internal technical capability.	Pressure to reduce cost and improve quality.
Predominant type	Frequent major changes in products.	Major process changes required by rising volume.	Incremental for product and process, with cumulative improvement in productivity and quality.

SOURCE: Adapted from Abernathy and Utterback (1978, p. 40).

(1) the character of the production process(es), (2) the diversity of the product line, and (3) the nature of innovation.[1] Table 3.1 describes several characteristics of the stages of development of a product. At the early stages, new products typically lack well-defined performance criteria, and market needs or process difficulties are approached through a variety of different product or equipment designs. Given a broad spectrum of possible designs, each embodying a fundamentally different technology, the product line is necessarily diverse. As a result, change is rapid and often alters the nature of the product itself. The production process, in turn, must be highly flexible, relatively labor intensive, and somewhat erratic in workflow.

At later stages of development, however, technological diversity gives way to standardization. Innovation, even if significant, alters only a small aspect of the basic product. Indeed, innovation at the mature-product stage is often difficult to perceive for any but the most knowledgeable engineers working on the project. Economies of scale guarantee a production process unlike the fluid "job shop" of the early years. Workflow is now rationalized, integrated, and linear; skilled labor is now replaced by highly specific "dedicated" equipment.

The development of technology from the fluid to the specific or mature state is initially a process of successive selection among competing design concepts; at the latter stages, it consists of refinements and extensions of concepts currently in use.[2] In identifying the nature of technical change associated with this pattern of evolution, it is helpful to distinguish between radical and incremental innovation. As used in this analysis, product innovation is labeled "radical" if it cannot be produced effectively in the existing production process. An incremental innovation, in contrast, utilizes the existing setup. The labels "radical" and "incremental" refer not to the change itself but to its impact on the production process.

It is essential in this context to distinguish between the general design concept and specific improvements in that concept through technical change. In automobile engines, for example, the V-8 gasoline engine was a general design concept that underwent a long series of improvements through innovations in materials and mechanical features. Such innovations are incremental; they refine and improve a general design concept that is currently in use. Radical innovation occurs with the introduction of a new approach or concept that cannot be produced effectively with the existing production process. A radical innovation need not be completely novel. Radical departures from existing concepts may have been known and available for some time but not used because of market preferences, relative prices, or technical problems.

The evolution from radical to incremental innovation is

characterized by a "technological hierarchy" within each of the various technical systems or components that make up the product in question.[3] This hierarchy of development arises out of technical and economic constraints that strongly influence the sequence of system developments. To use the engine as an example, the fundamental design choice seems to have been the type of fuel (which also implies external versus internal combustion). Once the gasoline engine achieved market dominance, other aspects of the engine (cylinder configuration, fuel delivery, materials) were successively chosen, developed, established, and then refined. The sequence is rarely of a rigid or linear sort. Many innovations come in bunches and interact with one another. Yet some appear to be of a more fundamental nature, affecting a significant number of cooperating features or aspects of the technical system; these require priority in development.

It is clear that a critical point in the transition from fluid infancy to standardized maturity is the development of a "dominant product design"--a synthesis of earlier innovations and design concepts that achieves significant market acceptance.[4] Both in components and in systems and overall product configuration, a dominant design permits standardization and economies of scale and, thus, introduces cost as a major aspect of competition.

THE U.S. EXPERIENCE

At first glance the automobile industry appears to be an exception to this process of development. A growing diversity in styling, model choice, and available options seems to belie any broad movement toward standardization. Appearances, however, are deceiving. Despite apparent diversity the underlying move in that direction has been pronounced. The pattern is well illustrated by the development of the gasoline engine.

Developments in Engine Technology

We have already noted the diversity of engine options available in the early days of the industry. Following the market's selection of gasoline over the electric and steam designs, technical change was focused on the development of cylinder configuration, mechanical efficiency, and materials. Though the basic combustion concept (internal) and fuel (gasoline) had been selected, there was a great deal of experimentation with other aspects of design. As Charles Sorenson, Ford's production manager in the Model T years, put it:

... it took four years and more to develop the [engine for the] Model T. Previous models [two-, four-, and six-cylinder] were the guinea pigs, one might say, for experimentation and development.[5]

The history of engine development at Ford is presented in Table 3.2. The table indicates the various cylinder configurations and the range of displacement and the number of different models in each category. Until 1970 two epochs are evident. The first lasted from 1910 until 1932 and was dominated by the four-cylinder IL-LH engine. The only other engine produced in that period was a V-LH 8 used in Lincolns. The second major epoch stretched from 1947 to 1970 and was the heyday of the V-OH 8. In this scheme, the period 1932 to 1942 was a transition era in which the V-LH 8 was joined by several V-LH 12 models in the Lincoln and by in-line, four- and six-cylinder versions. The V-OH 8 emerged as the dominant design, although several IL-OH 6 engines were available on small models.

At the same time that a single configuration achieved dominance, manufacturers offered an increasing range of size and performance options. Thus, from 1958 to 1970, Ford produced 15 different sizes of the basic V-OH 8 engine. Moreover, the basic engine was constantly refined and developed through the use of new materials and components. Yet from a manufacturing standpoint, and from the perspective of competitive rivalry, the engine offerings at Ford were highly standardized. Diversity in the size of the engine did not require diversity in process capability. Quite the opposite was true, because the dimension along which variation was introduced (cylinder size) was relatively easily accommodated in the same production process.[6] Likewise, the innovations that advanced engine capabilities preserved the competitiveness of the existing concept and extended its range of performance. Even though from a technical or engineering standpoint developments in such materials as grey cast iron may have been significant and even revolutionary, little change in basic manufacturing processes was required to implement a new material.

The standardization of the engine was intimately related to changes in the engine production process. Originally characterized by ill-structured tasks, highly skilled craftsmen, a job-shop workflow, and general-purpose equipment, the production of engines was transformed into a tightly integrated process utilizing operative skills, dedicated equipment, and much higher levels of automation. We refer not to the modern engine plant but to the engine plants of the late 1920s. In the case of Ford the surge in volume following the Model T both facilitated and made imperative the introduction of a less flexible, more specialized process capable of turning out a standardized product at lower and lower costs.

TABLE 3.2   A Summary of Ford Engine History, 1903-1974[a]

Engine Specifications

Number of Cylinders	Engine Configuration	1900-1910	1910-1922	1922-1932	1932-1942	1947-1958	1958-1970	1970-1974
2	OP	2 100.5-127						
4	IL	2 150-330						
4	IL-LH		1 177	2 177-200				
4	IL-OHC							3 98-140
6	IL	1 453						
6	IL-LH				1 226			
6	IL-OH					3 215-226	6 144-240	3 200-250
6	V-OH							1 171
8	V-LH			3 358-385	3 136-239			
8	V-OH					3 249-340 13 270-430	15 230-462	5 302-460
12	V-LH				5 267-448	1 292		

[a] Each cell contains the numbers of models and the range of CID produced at some point in the period.

NOTE: OP = opposing cylinders; IL = in line; LH = valve in head cylinder block; OH = overhead valve; OHC = overhead cam.

SOURCE: Adapted from Abernathy (1978, pp. 92-93).

Modern developments have accentuated the pattern established in the 1920s. Two trends have been dominant. The first was the continued development of automatic transfer machines that link operations without operator intervention. The growth of transfer capabilities preceded the second major trend: the development of higher levels of automation involving feedback control and machine self-correction. These advances have occurred within the context of a relatively stable product design and continued specialization or dedication of a particular production line to a particular engine.

The Body and the Assembly Plant

The process of technical development played out in the engine--establishment of dominant design, refinement, and extension--was repeated in the other major technical systems of the vehicle. Work on transmissions, bodies, and other components resulted in the development of preeminent concepts. Historical accounts make clear, however, that product technology in Ford's assembly plant (bodies, components, and the like) remained fluid and unstable far longer than was the case with engines.[7] A string of early innovations increased the scope and variety of assembly operations, among them: left-hand steering wheel (1908), steel running boards (1909), electric lights made standard (1915), baked enamel finishes by dipping (1917), starter available as an option (1920), and pyroxylin paint multicolors and closed steel bodies (1925).

In time, though, the pace of radical innovation diminished, giving way to the annual model change as the principal source of incremental innovation in body configuration. With the evolution of common body/frame "families" and the standardization of components within families, diversity among models has proven more and more a styling--and not a technological--reality. All that distinguishes most models from each other are appointments and trim; technological differences are embedded in component lines.

Variation in styling and similarity in technology thus offer double support for the U.S. automakers' traditional balance of marketing strategy with production efficiency. Differences among models most important in the showroom simply do not bulk large in the production process. Between 1949 and 1972, for example, the fraction of Ford assembly plants that produced only a single car (i.e., cars of a single wheelbase) rose from 6 to 35 percent.[8] And in those plants that produced two cars, the cars generally came from the same family. How different this is from the situation before World War II when many of Ford's 32 assembly plants were involved in the production of each Ford car.

TABLE 3.3   Ford Assembly Plant Equipment Characteristics, 1914-1974

Characteristic	1914	1927	1930-1936	1953-1955	1964	1970	1974
Transfer span	1	1	1	1	1	2	6
Group operations	1	1	1	144	144	90	130
Mechanization level	4	4	4	8	9	9	12

NOTE: *Transfer span* is defined as the maximum number of different machine tool operations that are linked together and whose operation is controlled by an automated transfer machine. *Group operations* is the maximum number of different operations performed simultaneously. *Automation level* is an index of the maximum level of automation achieved in a machine tool; based on a scale developed by Bright; for example, a machine at level 12 changes speed, position, direction according to measurement signal; level 4 includes simple power tools under hand control.

SOURCE: Abernathy (1978, p. 134).

Modern assembly operations are highly mechanized, integrated, automated, and specialized. The data in Table 3.3, which presents three characteristics of process technology at Ford during the period 1914-1974, document this point very clearly. When Ford changed over from the Model T to the Model A in 1925, all production facilities had to be closed down for a period of nine months, which effectively abandoned market leadership to General Motors (GM). Though the conversion wrought great changes in the engine and components plants (many machines were scrapped altogether; 4500 new ones were bought and over 50 percent of the machine tools were rebuilt), the assembly plants were only minimally affected.

By contrast, conversion of a modern assembly plant to a new vehicle family takes months of planning and several months of retooling. Unlike their highly flexible precursors, the assembly plants of the mid-1970s were highly specialized and capital intensive.

Two additional points are worth making about the development of the automobile at Ford. Though we have spoken mostly about engines and bodies, we could just as easily have spoken about transmissions or other major components. On balance the market views the car as a whole, and the evolutionary picture we have drawn applies to the whole car as well as to its individual systems or components.

The Model T, for example, was a dominant design for almost 20 years. It was designed to capture the "basic transportation" market and embodied a synthesis of major advances designed to reduce weight, toughen construction, increase reliability, and lower cost. The development of closed steel bodies in the 1920s changed the character of the automobile. "Basic transportation" was replaced by roominess, comfort, and smoothness of ride as principal design criteria. Developments along these lines led to the all-purpose road cruiser, which dominated the U.S. market from 1948 to 1970. In a functional sense, the designs of the major manufacturers in that era were quite similar. The dominant overall configuration included a large V-8, water-cooled, front-mounted gasoline engine, with rear-wheel drive, automatic transmission, and a comfortable roomy interior.

Throughout the era of the all-purpose road cruiser, improvements in technology (as opposed to, say, the great changes in sheet metal usage and appearance) have by and large been the result of incremental and not radical innovation. Nor is this pattern of development limited to Ford. By the early 1970s all of the major U.S. manufacturers had undergone a comparable evolution through the stages outlined above. All of them became the purveyors of a standardized product and the masters of a specialized process technology.

CONTRASTS IN THE EUROPEAN EXPERIENCE

When compared with European developments the evolution of the automobile in the United States reflects both a distinctive technological thrust and a particular mode of competition. Though innovation in each technical system was not the result of a coordinated development effort, the character of the changes introduced were related through the driving force of consumer preference and market competition.

Almost from its beginning the U.S. industry was oriented toward the mass market and the provision of a comfortable, reliable, general-purpose vehicle easy to operate and to maintain. Major developments in the technical systems that achieved market dominance were those that reduced costs, increased comfort, and eased operation. Model changes in the pre-World War II years were more important competitively than in later years when designs stablized, but the market in the 1920s and 1930s did not demand continuing advances in the technical sophistication of the product. Indeed, sophisticated engineering features, or advanced technical changes that departed from the main lines of development, often met with market failure.

Rather than advanced designs and engineering, demand centered on costs and styling and acceptable levels of performance. As was noted in Chapter 2, the success of GM's strategy seems to confirm the secondary role of evident technical advance in competition. GM avoided competition on the basis of advanced technology and adopted an approach emphasizing incremental change, acceptable designs, and broad product-line policy to meet market needs.

The contrasts with European developments are instructive. Although the industry began in Europe, it grew far more vigorously in the United States.[9] By the 1920s, 1 out of every 5 Americans owned an automobile; in Germany, only 1 out of 56. In comparison to their U.S. counterparts, European drivers were more sophisticated. They were attracted to features that required and enhanced driving skill. It was not a mass market. In fact, not until the 1950s did car ownership in Europe become genuinely widespread. Beyond such differences in timing, systematic government policy after World War I helped distinguish the European industry from the American. Various tax and regulatory policies defined the market and shaped its development in each of the producing European nations. In Britain, for example, there was a horsepower tax, which strongly influenced the development of a small-bore, long-stroke engine. Then, too, high fuel costs-- also, to some extent, a reflection of government tax policy-- placed an early premium on fuel efficiency. The result was a path of technical development that emphasized vehicle performance.

Beyond tax and regulatory measures, government policy on transnational trade had a profound influence on the growth of distinctly national markets. Surrounded by relatively high tariff barriers (e.g., the French had a 90 percent tariff in 1931) and motivated by distinctive domestic tastes and preferences, the European auto firms developed sharply differentiated products with distinctive national characteristics. Thus, the major German firms produced cars that were quite different from those developed by the French.

Within the context of broadly similar national technology, the particular firms in each country developed products and systems that were different. While BMW products, for example, were more closely related to those of Audi and Mercedes than they were to those of Peugeot or Renault, their design and technical features were distinctly different from their domestic competitors. Both among countries and among firms, the phenomenon of a "dominant design" failed to emerge in most of the major technical systems. In engines, suspensions, drive trains, fuel delivery, and so forth, a diversity of technology characterized the European market.

The absence of a dominant design and the consequent diversity in automotive technology in Europe seem to have been a result of the nature of competition. It is true that government trade policy had a strong bearing on the growth of national markets, but diversity has persisted long after the European Common Market was established.

Table 3.4 provides examples of the kind of distinctions industry experts use to characterize products of the major producing countries in Europe. These differences reflect a long tradition of technological development in each country that has been preserved despite greatly increased inter-European trade. It appears that preferences and tastes remain sufficiently diverse to support a range of designs and technical options.

Moreover, the search for competitive advantage demands it. The European emphasis on vehicle performance created opportunities for competitive advantage through nonincremental innovation and advanced engineering. The industry originated less in the demand for basic transportation and more in the search for high-performance luxury vehicles. Cost was far less important; technical performance was essential. In this sense the European industry retained a level of diversity in design approaches more characteristic of the fluid stage of U.S. development.

In this context it is not surprising that the European subsidiaries of U.S. companies have been operated as separate businesses. This is true not only in terms of manufacturing, product policy, and market strategy but more importantly in terms of organizations, systems, and personnel. The changes in the U.S. market in recent years have prompted concerted effort to bring

TABLE 3.4   National Characteristics in European Automobiles

Country	Major Producers	Characteristics
France	Renault, Peugeot-Citroen	Soft ride; low performance, highly idosyncratic styling.
West Germany	VW-Audi, Opel, BMW, Mercedes, Ford	Quick acceleration; firm ride; high performance/high speed.
Italy	Fiat, Alfa-Romeo	High revolutions; specializing in small sporty cars.
Sweden	Volvo	Large, heavy cars; safety orientation; distinctive styling, boxy.

SOURCE: Discussions with industry observers.

the U.S. and European pieces of the U.S. domestic producers closer together.

We have argued that technical diversity and national identity characterized the European market from its inception and that they have persisted in the face of large trade flows. There are indications, however, that the quite different U.S. pattern of development is present in the low-cost segment of the market. Within the last few years, for example, the major manufacturers have developed low-priced cars with very similar technical configurations. Indeed, the Renault R5 with its boxlike exterior, front-wheel-drive, four-cylinder engine, and 4- to 5-gear manual transmission seems to have established a dominant design in what might be called the "econobox" segments. Ford (Fiesta), Fiat (Strada), VW (Rabbit), and GM (Kadett) have all offered products with a similar design approach.

The appearance of standardization in the low-priced segment reflects the efficiency orientation of consumers in this market. Performance in terms of handling, power, and so forth are less critical than low-cost operation and efficient use of space. Efficiency and cost also seem to have played a role in the decision of Ford of Europe to break away from nationally based designs. The desire to rationalize its European operation led Ford to develop a truly European product line and to coordinate its European production facilities. As in the United States, Ford sought to decrease the underlying technological diversity of its European products at the same time that it increased their variety in styling and appointments. Though Ford offered, say, four engine types within a given model, it kept those engines common across several model lines. While specializing engine production by plant, Ford could thus retain a real measure of choice on the showroom floor.[10]

Furthermore, European-wide sourcing of components allowed

plants to be dedicated to a smaller number of products than before. The Fiesta plant that Ford built in Valencia, Spain, specializes entirely in the production of Fiestas. It is, of course, less flexible than the older plants, which had to accommodate several different models, but it has been able to achieve far higher levels of automation and integration.[11]

The move to European-wide sourcing and increased common-ality is also apparent at Volkswagen (VW). Commonality and standardization have been introduced to such an extent that VW's whole European product line uses only one automatic transmission. The eight models in its line (and the many variations within those models) require only five basic platforms and four basic engines. While the available evidence suggests that GM's subsidiaries have developed a similar approach to design and sourcing, other major producers have continued in traditional patterns.

A degree of standardization in the lower-priced segments has been a factor in the recent penetration of the Japanese into the European markets. While not as technically sophisticated as some of the leading European firms, the Japanese have developed very reliable vehicles of acceptable function and styling and are selling them at prices below comparable European products. In contrast to the United States where the Japanese price their vehicles above the market, in Europe the Japanese have used a penetration pricing policy. This approach has been highly successful in those segments where efficiency and low cost are paramount.

TECHNOLOGY CONVERGENCE:
IMPLICATIONS FOR COMPETITION AND ORGANIZATION

The diversity of product technology in Europe underscores the intricate connection between the character of competition and the pattern of technological innovation. With much less emphasis on product performance in the U.S. competitive arena, particularly after 1945, technology assumed a neutral role in the fight for competitive advantage. The very notion of convergence in product and process characteristics across firms implies a similar evolution of technology and organization within firms. The patterns are not identical either in specific forms or in their timing, but in its fundamental characteristic the evolutionary process has signifi-cant commonalities. Evidence for this proposition is provided by the pattern of diffusion of several major innovations. Figure 3.1 presents diffusion data for a few significant technological developments from 1910 to 1974. The rapid diffusion evident in the data suggests that the productive units in the major firms were at similar stages of development when the innovations were introduced.

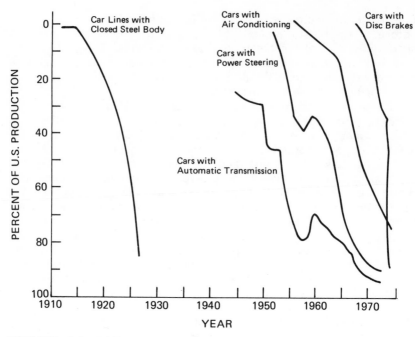

FIGURE 3.1 Diffusion of selected innovations. (Adapted from Abernathy, 1978.)

The almost complete diffusion of the innovations in Figure 3.1 suggests that any competitive advantage accruing to the innovator was shortlived; what was initially a unique feature available on a limited basis became widespread, even standard equipment on all cars. Where competitors are at similar stages of development, and where development has proceeded through a particular sequence of dominant designs, technology becomes competitively less significant. Because all firms have evolved in a similar fashion, no single firm can sustain a competitive advantage through incremental product innovation. Rapid replication by competitors quickly eliminates any gains. Under these circumstances the incentive for significant product innovation is diminished. Innovation occurs, but as we have seen it is increasingly incremental, defensive, and invisible.

This is not to argue that the innovative process in a mature or maturing industry is not a significant factor in any given firm's competitive survival. Clearly the product does evolve and change; the production process becomes increasingly more productive. Without refinements in existing design concepts a firm will fall by the wayside. What is important however is that innovation is incremental and slowly cumulative in its impact. It is critical for survival but not for competitive advantage.

This changing pattern of innovation as a product matures is accompanied by an evolution in the capabilities of the firm as an organization. The organizational changes are likely to be complex, but a few key stylized facets of development will serve to indicate the basic pattern of evolution.

As far as technology is concerned, the key competitive task for the maturing firm is the steady refinement of design concepts currently in use. This fact conditions the kinds of technical changes made, the character of technical and human resources the firm acquires, and even the origin of improvements. Innovations of a radical sort are destructive of existing capital and generally highly risky in both the market and technical senses. A sweeping shift to totally new design concepts requires an entrepreneurial thrust both in its technical development and in its commercial application. In contrast, an organization with a dominant orientation toward mass production of a mature product, economies of scale, and incremental innovation must place far greater emphasis on cost control and coordination. Entrepreneurship in such a setting may be quite dysfunctional; where it exists, it is likely to be organizationally separated from the core activities of the firm. Rather than brilliant but risky technologies, the firm oriented toward incremental innovation will emphasize engineering applications that push existing in-use technologies to their limits.

With an increasingly complex process the successful firm in a maturing industry is likely to evolve an organization and a management team that excels at coordination and control.[12] As the production process becomes increasingly capital intensive and complex, there is likely to be greater specialization of tasks both for workers and managers and an increasingly hierarchical organization.

With competitive emphasis on production, costs, and incremental change, the organization adapts to support that thrust. Firms in the auto industry seem to fit this pattern of organizational evolution quite well. If the auto companies excel at anything, it is in the efficient operation of a highly complex production process. Indeed, if one were to ask what does the auto industry do well, coordination and control--essentially a cost emphasis through exploitation of economies of scale--would be high on the list. It is important to see, however, that the organization's capabilities in other dimensions--rapid innovation, for example--may be more limited.

The implication for responses in the current crisis are clear: Should major changes in strategic emphasis be required, successful adaptation will involve a fundamental organizational transformation as well as changes in characteristics of products and technologies. The changes required will apply not only to design and engineering but also to the relationship between these functions and operations and to top management. If innovation becomes a more significant factor, a closer integration of R&D and the marketplace will be required. Design targets and the discipline imposed may well change. Competition will turn more on the ability to bring new ideas into operation than on the bureaucratic control of the cost of a standarized product.

NOTES

1. The basic notion of process and product evolution has been developed in a series of articles by Abernathy and Townsend (1975), Utterback (1974), and Utterback and Abernathy (1975). For an extension and application to the auto industry, see Abernathy (1978).

2. The notion of a "design concept" in this context was introduced in Abernathy (1978), p. 54.

3. The technical hierarchy referred to here has been discussed in Abernathy (1978), pp. 20, 62-65.

4. For a discussion of the concept of "dominant design," see Abernathy (1978), p. 57, and Abernathy and Utterback (1978), p. 46.

5. Sorensen (1956), p. 102.

6. Abernathy (1978), pp. 95-97.

7. For evidence on this point, see Abernathy (1978), pp. 114-143.

8. Abernathy (1978), pp. 131-132.

9. For a wealth of historical data on the international automotive industry, see Wilkins (1980).

10. See Doz (1979) on these matters.

11. Ibid., pp. 20-22.

12. The evolution of organization in the auto industry has been examined in Abernathy (1978) and Chandler (1964).

4

International Competition:
Trade Flows and Industry Structure

Until the last decade the evolution of the U.S. automobile industry was largely determined by political and economic forces specific to the North American continent. The extent of the U.S. market and its production base had long sustained a largely self-contained industry. In the last few years, however, the relevant industry boundaries have expanded dramatically.

This chapter documents the two major changes visible in this internationalization of the U.S. auto industry's market and product. The first has to do with the volume and pattern of world trade in automobiles during the last 75 years; the second, with the number of competitors and the structure of the world market. Both developments have, of course, been shaped by government policy, by shifts in consumer preferences, and by the evolution of product and process technology. These forces are particularly well illustrated by the history of the small car in the U.S. market, a history we examine in concluding the chapter.

PATTERNS OF WORLD TRADE IN AUTOS

In the very early years of the industry the U.S. component of world trade was dominanted by a protective U.S. tariff and, at least for U.S. producers, high transportation costs.[1] Until 1913 the U.S. industry had developed into the world's largest automotive industry behind the protection of a 45 percent ad valorem tariff. By contrast, European countries had a relatively open tariff policy in the pre-World War I era but maintained various horsepower and other use taxes that together with high transportation costs meant that significant U.S. penetration of foreign markets, particularly for low-priced products, could not be accomplished by exports. Thus, even at a time when the British had no tariff (1911-1912), Ford opened production facilities in the United Kingdom. Differences in consumer tastes and in technology also influenced trade

TABLE 4.1 A Comparison of Automobile Trade Flows Among Major Producing Countries in 1955 and 1970

Importer	Exporter							Imports[a]	Total Cars Registered	Imports as a Percentage of Total Cars Registered
	United States	Canada	Japan	West Germany	France	Italy	United Kingdom			
United States	—	5	N/A	37,959	3,803	324	19,464	61,555	7,682,630	0.79
	—	696,730	353,779	654,937	41,467	45,583	76,891	1,927,120	8,388,204	14.70[b]
Canada	27,522	—	N/A	6,628	178	9	14,155	48,492	363,697	11.76
	245,616	—	67,931	37,420	17,709	4,602	21,640	405,176	636,206	63.69
Japan	765	4	—	428	3,279	6	5,040	9,522	13,354	41.62
	5,058	5	—	9,560	140	632	1,781	17,766	2,379,129	0.75
West Germany	983	11	N/A	—	4,061	12,610	3,852	21,517	387,980	5.25
	2,476	4	404	—	229,467	173,509	14,977	426,638	2,107,123	20.25
France	4,742	20	N/A	2,364	—	732	2,763	12,621	426,713	2.87
	394	13	1,450	85,122	—	88,096	5,856	181,428	1,296,628	13.99
Italy	314	2	N/A	1,764	645	—	660	3,385	161,583	2.05
	159	14	883	180,898	139,676	—	53,472	375,474	1,363,594	27.54
United Kingdom	700	149	N/A	5,694	4,231	1,908	—	12,682	529,560	2.34
	434	20	5,160	38,067	47,523	30,695	—	153,809	1,076,865	14.28
TOTAL								169,774	9,565,417	1.77
								3,487,405	17,247,749	20.20

[a] Importing countries include the United Kingdom, the United States, Canada, France, Italy, West Germany, Sweden, and Japan.
[b] Excludes imports from Canada.

NOTE: The top entry in each cell is the number of vehicles imported (exported) in 1955; the bottom entry is the number for 1970.

SOURCE: Society of Motor Manufacturers and Traders Motor Industry of Great Britain.

flows. As we noted in Chapter 3, the car in the United States was more of an all-purpose workhorse; the more sophisticated European motorist demanded a different car. There was a pronounced aversion in the United States to technical features that required driver skill, such as precision gear shifting. Compared with Europe, where horsepower taxes and high gasoline taxes dictated a performance-oriented vehicle, American manufacturing sought to emphasize driving simplicity, lower costs, and engine flexibility. Although U.S. producers remained the world's leading exporter for the next 40 years, the numbers of cars exported were small. Substantial U.S. involvement abroad occurred through direct investment.

In the years following World War I, as transportation costs became a less significant barrier, government trade policy grew in significance. The British introduced a 33-1/3 percent tariff in 1915 and later added a horsepower tax, which significantly shaped the development of the British industry. Similarly, the French, acting to protect their domestic industry, set a tariff of 45 percent in 1922, raising it to 90 percent in 1931. West Germany used a combination of tariffs, foreign exchange restrictions, and local content requirements to provide an effective measure of protection. In the face of such restrictions on trade, Ford and GM acquired or established a significant number of manufacturing facilities in the major European countries; by 1929 the facilities of the two companies numbered 68.

On the eve of World War II, direct international trade in automobiles was insignificant. European producers were insulated from foreign competition and operated in protected national markets. The U.S. producers also were protected, not by government policy it is true, but by the competitive strength of the domestic industry. Dealer and service network played an important role in marketing, making penetration by an importer difficult. Furthermore, the products in demand in the United States were far different from those being produced in Europe, where taxes on gasoline and horsepower led to production of automobiles that offered more performance from engines and fuel but at the sacrifice of popular prices, vehicle utility, and the development of mass markets. Thus, barriers to imports were inherent in the nature of the U.S. industry and its unique market demand, although from a policy standpoint the United States was an open market.

Two principal developments characterized international trade in the immediate postwar years. The first was the openness of the U.S. market to growing import penetration in response to changes in consumer demand; the second was the formation of the European Economic Community (EEC) and the resultant growth of inter-European trade. Both developments are evident in Table 4.1, which compares trade flows of 1955 and 1970.

TABLE 4.2   A Comparison of Automobile Trade Flows Among Major Producing Countries: Year Ending December 1978

Importer	Exporter							Imports^a	Total Cars Registered	Imports as a Percentage of Total Cars Registered
	United States	Canada	Japan	West Germany	France	Italy	United Kingdom			
United States	—	783,755	1,408,669	423,096	33,178	72,246	55,134	2,839,330	10,946,000	18.8^b
Canada	542,341	—	119,568	53,694	13,143	3,999	7,015	745,017	958,414	77.73
Japan	13,182	21	—	30,159	1,611	1,286	2,830	51,113	2,856,710	1.79
West Germany	14,110	1,385	119,287	—	245,842	116,790	17,688	565,939	2,663,754	21.25
France	988	48	32,236	135,572	—	90,443	14,981	493,604	1,944,986	25.38
Italy	104	9	1,138	127,242	238,966	—	7,660	440,677	1,215,731	36.25
United Kingdom	1,389	180	143,116	206,173	158,255	97,948	—	797,807	1,591,941	50.12

^a Importing countries include the United Kingdom, the United States, Canada, France, Italy, West Germany, Sweden, and Japan.
^b Excludes imports from Canada.

SOURCE: Society of Motor Manufacturers and Traders Motor Industry of Great Britain.

Note, first, the tremendous increase in the volume of trade between 1955 and 1970. For the major producing countries taken together, the ratio of imports to total production rose from 1.8 to 20.2 percent. The share of imports rose in each country, but the rise was most dramatic in the United States. In 1955 imports held less than 1 percent of the U.S. market, with products from Germany accounting for almost two-thirds of this amount. By 1970 imports had taken 15 percent of the U.S. domestic market. German products (principally VW) continued to lead the importers but were under significant challenge from the Japanese. Japan emerged as a major factor in the world market only in the 1960s, almost entirely through its sales to the United States. In 1970 Japanese exports to Europe were trivial.

Although the Europeans were not buying Japanese products in 1970, the Europeans were involved in significant trade among themselves. Among the three major EEC producers--West Germany, France, and Italy--there had been, as in the United States, an obvious increase in the volume of trade. But in contrast to its leading position in the U.S. market, Germany was a net importer within the EEC. So for that matter was Italy; France, however, was still a net exporter of automobiles. Indeed, the imports share of the French market in 1970 was slightly below that found in the United Kingdom, which was not a member of the EEC at that time.

Patterns of trade after the first oil shock in 1973 remained about the same, even though the pace of developments had quickened. Table 4.2 presents data on trade flows in 1978. By the end of the 1970s, Japanese producers were the dominant exporters to the United States. The extent of their increased market penetration is striking. The total U.S. market grew by 30 percent from 1970 to 1978, but the Japanese share increased over fourfold. This time, however, the rise of Japanese imports was not limited to the United States. In West Germany and the United Kingdom, the Japanese penetration was, if anything, more rapid still. From virtually zero percent in 1970, Japanese products by 1978 came to hold 9 percent of the British market and 4.5 percent of the West German. Since imports from outside the EEC accounted for only 4.2 percent of all new cars registered in the EEC (excluding the East Bloc), it is clear that Japan is the dominant force in opening the EEC to outside penetration.

These developments, moreover, do not appear to be transitory, for recent evidence suggests that the Japanese producers have continued to make significant inroads into European markets, particularly in Britain and West Germany. Their entry into Italy, however, has been barred by a restrictive quota, and such policies may well become more prevalent, especially if local demand shrinks in the face of expanding capacity.

It was not only the Japanese who entered the British market. Membership in the EEC opened Britain to a virtual flood of continental products. Imports held 14 percent of the British market in 1970; their share in 1978 had grown to just over 50 percent. West Germany, followed by France, lead the EEC importers to the United Kingdom, and, as a result, Germany has joined France as a net exporter within the EEC.

The data on international trade underscore the importance of government policy in shaping the extent and patterns of trade. From the earliest days of the world industry, tariffs, taxes, quotas, and regulations have been used in one form or another to influence the location of production and the volume of trade. The most prominent example of government intervention in the industry has been the export promotion policies of the government of Japan. Although reduced somewhat in the last decade, the policy of the Japanese government has been to protect the domestic industry from import competition, while encouraging the growth of exports through a variety of subsidies. The most obvious polar extremes in policy terms have been the situation in the United States, where the domestic industry has been insulated from foreign competition by the barriers inherent in the U.S. market and where government policy has been oriented toward the free flow of products.

The penetration of the Japanese into Europe and the United States and the apparent slower growth of demand for automobiles under conditions of worldwide recession have created pressure for governments around the world to protect their domestic auto industries. Of course, the protection and nurture of a domestic auto industry is not a new event. Apart from the major producing countries, all of whom have operated behind tariff barriers at one time or another, developing countries have made extensive use of government policy to encourage the growth of automobile production.

Trade involving the less-developed and developing countries, though of little moment before the 1973 oil crisis, has become more significant. Patterns of government policy, much as in the early years of the auto industry, are reflected in patterns of trade. Table 4.3 provides a summary of trade policies of the major producing countries and some representative developing nations. Many of the developing countries have established barriers to imports in order to encourage and protect domestic production. Primarily through the use of local content requirements, such countries as Mexico, Brazil, and Korea have developed domestic production bases that are beginning to compete actively in the world market. Particularly in terms of component manufacture, the developing countries are increasing their exports. These emerging patterns of trade depend heavily on the absence of restrictions in the major markets in the world, particularly in

TABLE 4.3   Policies Affecting Trade in Automobiles in Selected
Countries, 1980

Country	Tariffs	Special Vehicle Taxes	Local Content Regulation
United States	2.9 percent	None	Local content applies to CAFE standards.
West Germany	10.9 percent (EEC)	Annual road tax based on cylinder capacity.	None
United Kingdom	10.9 percent (EEC)	None	None
France	10.9 percent (EEC)	Annual vignette tax based on age or "fiscal" horse-power.	None
Italy	10.9 percent (EEC); quotas on Far Eastern products (Japan 1980 quota: 2,200 vehicles).	Annual use tax based on engine capacity; annual certification tax related to size.	None
Japan	None	Engine displacement shadow area.	None
Brazil	185-205 percent; imports currently embargoed.	None	Negotiated individually
Spain	68 percent (non-EEC/EITA)	Luxury tax based on horsepower.	55 percent
South Korea	80 percent; import license required.	—	Varies (by type of car) from 62 to 94 percent.

SOURCE: Mark B. Fuller, "Note on the Auto Sector Policies of France, Germany, and Japan." (Harvard Business School Case Services, 1981).

the United States. Thus, if open markets prevail, it is not unreasonable to expect production from lower-cost areas to increase in significance in world trade.

THE STRUCTURE OF MARKETS AND COMPETITION

The evolution of technology, the rapid growth of world trade, and the emergence of mass consumption in Europe and Japan in the postwar era have had a profound effect on the number, strength, and strategies of competitors in the world auto industry. Before reviewing these changes in industry structure, it will be useful to sketch out the conceptual framework that underlies our interpretation of them.

In an industry where production hinges on a relatively stable and makeable product technology and where the manufacturing process is capital intensive and offers significant economies of scale, periods of growth in market size create opportunities for producers to offer an increasing variety of products at or below the cost of the old product mix.[2] If the technology of production is well understood and procurable (e.g., embodied in capital equipment that can be purchased or in human skills that can be readily hired), then in addition to existing firms it would be expected that opportunities for identifying and serving new market niches will be exploited either by firms on the fringe of the industry or by new firms. Unless barriers to entry are substantial, growth in overall market size will not be accompanied by a proportionate growth in the market shares of the leading firms. The large firms will certainly grow, but some of the market increase will be absorbed by those new or fringe firms offering additional variety or serving special market segments. Thus, seen in historical perspective, developments that extend economies of scale and add to the maturity or stability of technology will result in a deconcentration of the market and an increase in the effective number of competitors.

With major shifts in process technology that underpin product advance, however, the trend toward deconcentration may experience a reversal. The reason for this lies in the "lumpiness" of new technologies and in the problem of procuring and mastering the new process. When advances in process technology depart significantly from existing approaches, they usually involve large capital outlays and the development of hard-to-acquire skills on the part of workers and particularly management itself. This is true even though the new process may offer significant reductions in the variable cost of production or important new product features. Significant increases in volume are needed to warrant their introduction. Thus, growth in market size over time may create an incentive for the introduction of new, radically different high-volume techniques.

Such introduction is unlikely to be smooth in its effect on market shares. In the first place, an innovating firm may need additional volume to justify the change. Perhaps more important, the innovation is likely to create a competitive advantage that draws customers away from the fringe or newer producer. The result is to increase concentration and to reduce the effective number of competitors, at least until further growth in the market and renewed stability and diffusion of the technology allow the emergence of smaller, more specialized producers.

It is important to distinguish the connection between capital-intensive process innovation and market structure as outlined here and the relationship between firm size and technological innovation of a more general sort. It has often been the case, particularly in the very earliest phases of an industry's development, that innovation in products has been associated with the entry of small firms. When the product is changing rapidly, when alternative product technologies vie for market dominance, production processes are typically flexible and relatively less capital intensive. In these circumstances, product innovation can be a source of competitive advantage, and the innovating firm need not be particularly large nor command a significant share of the market to survive, at least for some time. It is only when product technologies stabilize and economies of scale become of critical importance that the important connection between concentration and process innovation emerges.

Concentration in the United States

Figure 4.1 illustrates the historical patterns of concentration in the U.S. auto industry.[3] (The measure of market structure used here is the number of equivalent firms--that is, roughly speaking, the number of firms that would populate the industry if all had the same market share as the larger firms. Technically, it is the inverse of the Herfindahl index, defined as the sum of the squared market shares.) Prior to 1955, three patterns of concentration are evident. The first and most radical changes followed the introduction of the process innovations associated with the Model T (1908-1919); the second occurred with the introduction of closed steel bodies (1924-1926); and the third resulted from the widespread introduction of automation in transfer lines in stamping, in engine production, and so forth (1948-1954). Each episode marked the introduction of capital-intensive technology that significantly altered returns to scale, and each was followed by a steady increase in the number of equivalent firms.

Two additional patterns of concentration have appeared since 1955. In the early 1960s the introduction of compact cars produced with automated equipment stimulated a brief period of deconcentration that lasted until the mid-1970s, when the move to downsizing produced another period of concentration.

Changes in World Market Structure

The ebb and flow of concentration has occurred against the backdrop of a long-term decline in concentration, which began in the

60

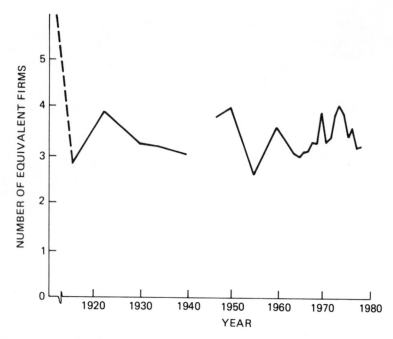

FIGURE 4.1 Trends in U.S. automobile market structure. (Data for 1900-1965 from Abernathy, 1978; data for 1966-1979 calculated from data in Ward's Automotive Year Book, various years.)

World War I era after the first major shakeout in the industry had established Ford as the dominant firm. This long-term trend is consistent with the general evolution of technology and competition in the industry and, although different in its specifics, is consistent with developments in the world market. The pattern of change in the number of equivalent firms worldwide since 1950 is presented in Figure 4.2.

In 1950 the number of equivalent producers worldwide was less than five, for the market was dominated by U.S. manufacturers and their subsidiaries. The growth of the market since 1950 and the emergence of several large producers in Europe and Japan are reflected in the long pattern of deconcentration in world markets until 1975. Since 1975, however, the world market has become more concentrated.

Table 4.4, which compares world market-share data for 1965, 1975, and 1979, presents evidence broadly consistent with the previously discussed trends in international trade. The Japanese have become the principal source of increased competition in the

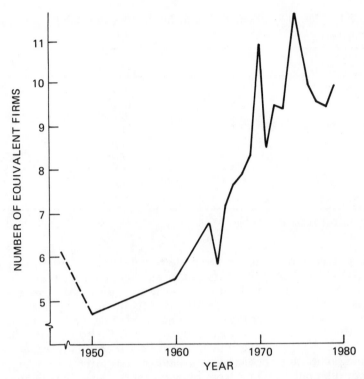

FIGURE 4.2 Trends in world automobile market structure. (Data for beginning to 1965: Vernon, 1977; 1965-1979: calculated from data in World Motor Vehicle Data, Motor Vehicle Manufacturers Association, 1980.)

world market. From 3.8 percent in 1964, Toyota and Nissan have increased their world market share to 13.1 percent, ranking fourth and fifth, respectively. The rise of the EEC also has had a major impact, with the most significant gains accruing to the French producers. In the last four years, GM has made significant advances, largely as a result of changes in the domestic market.

Taken together, the evidence from the U.S. and world markets is consistent with the pattern of interaction between share growth and technology discussed above. The internationalization of the auto industry has seen a long-term trend toward increased competition; both the number of competitors and their relative strengths have increased. The evident concentration in the last few years is a response to worldwide market shifts that have required significant capital outlays for new technologies and new modes of competition.

TABLE 4.4   Market Shares in the World Automotive Industry
(percentage of)

Producer	1965	1975	1979
General Motors	30.9	19.0	21.8
Ford	19.6	12.4	12.4
Chrysler	9.6	7.8	3.4
Volkswagen	7.3	4.5	5.3
Renault	3.0	5.3	4.6
Peugeot	1.4	3.0	8.3
Fiat	5.3	4.6	4.2
Toyota	2.5	7.3	7.4
Nissan	1.3	6.7	5.7

SOURCE: *World Motor Vehicle Data,* Motor Vehicle Manufacturers Association, 1980.

The patterns of concentration sketched out in Table 4.4 and Figure 4.2 reflect developments in the number and relative size of competitors in the world auto market, but they fail to reveal important changes in the character of interfirm rivalry and competition in the last several years. A number of developments have blurred the distinctions traditionally used to distinguish one firm from another. Firms have moved vigorously to gain advantages of scale and competing technology through various interfirm arrangements; agreements ranging from outright mergers to joint ventures have changed the traditional boundaries of many firms. More often than not these arrangements have been realized through the encouragement of host governments in the case of foreign firms. Europe has been the center of much of this activity, but U.S. and Japanese firms have also been involved at one level or another. Figure 4.3 presents a schematic representation of the equity relationship linking automakers around the world.[5]

Mergers and Acquisitions

In the history of the industry, mergers and consolidations have played a major role in shaping the size and number of competitors, particularly in Europe. Although infrequent in the United States, mergers between significant domestic producers have been witnessed by all the major European-producing countries. British Leyland in Britain, Peugeot-Citroen-Talbot in France, VW-Audi-NSU in West Germany, and Fiat-Lancia in Italy are the prime examples. The incentives for merger are familiar from our discussion of international trade. As with Ford of Europe in the mid-1960s, a desire to achieve economies of scale, rationalize

63

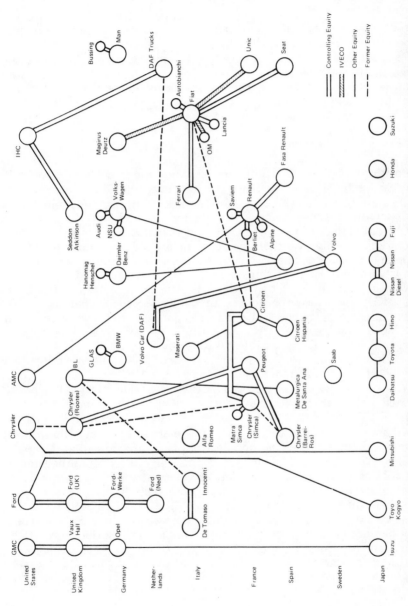

FIGURE 4.3 Equity relationships among selected automakers. (From Salter and Fuller, 1980.)

dealer networks, and exploit the potential for specialization and integration all play a role. The pressure for mergers is not restricted to national boundaries. The recent purchase by Peugeot of Chrysler-Europe (now called Talbot) is indicative of the desire for a European-wide system of production and distribution. Political pressures, however, often make such moves difficult. Whether the Peugeot case will be repeated in Europe is uncertain given the political realities, but the pressures arising out of increasing economies of scale and rapidly maturing domestic markets are unlikely to abate soon.

Marketing and Production Joint Ventures

Politics may stop outright mergers, but the underlying economic pressure for alliance is likely to spill over into other areas. One area of business where a joint arrangement can be valuable is distribution and service. The dealer network has long been a critical element in competition, and firms desiring to enter a new market or improve their position in an old one have found the presence or absence of a dealer network to be vital. Thus, it is no surprise that Renault sought access to an established network by working out an agreement with the American Motors Corporation (AMC). In return for access to dealers, Renault has agreed to provide capital to AMC and may design a car to be produced in an AMC production facility. In addition, recent arrangements have paved the way for a possible takeover by Renault.

Joint production arrangements are prevalent in the world market, involving most of the major producers. Table 4.5 provides a partial account of production arrangements worldwide. It seems clear that traditional notions of interfirm competition may have to be modified in light of these developments. If significant alliances develop short of outright merger or acquisition, there may be fewer effective competitors than are immediately apparent. The press for scale economies may force consolidation in practice if not in name, so that products of the cooperating firms, at least in technical aspects, become less distinctive and more standardized. Yet alliances may also preserve competition if they permit otherwise nonviable competitors to continue to compete in other areas. Firms cooperating in engine production may differentiate products in other dimensions. Joint ventures could result in the preservation of a major "second tier" of firms offering competition to the worldwide full-line producers. These considerations do little, however, to alter our basic conclusion regarding the forces shaping long-term trends in industry structure. If anything they reinforce the notion that stability in technology enhances competition, based on price and economies of scale, while process

TABLE 4.5   Examples of Joint Production Arrangements in the World
Automobile Market

Companies	Arrangements
Honda-British Leyland	Reciprocal marketing of vehicles; assembly of Honda products in British Leyland facilities.
Renault-Peugeot	Development and production of engines; (Français de Mécanique) development and production of gear boxes and axles (STA).
Volvo-Peugeot-Renault	Development and production of new-size cylinder engine.
Lancia-Saab	Production of new passenger car.
Ford-Renault	Assembly of Ford vehicles in Australia.
Alfa Romeo-Fiat	Assembly of reciprocal supply of components.

SOURCE: Salter and Fuller (1980).

changes and innovation, particularly of the capital-intensive
variety, are associated with periods of concentration.

THE FORTUNES OF THE SMALL CAR IN THE U.S. MARKET

In the auto market of 1980 the small car dominated sales. The
dominance of the small car has been a fact of life in most countries since the industry's birth. But in the United States the
recent rapid shift to smaller cars is nothing short of a revolution.
Until the mid- to late 1970s, the small car in the United States
was a specialized niche--a fairly good sized niche, but a niche
nonetheless--in the overall U.S. market. It was a segment traditionally filled by foreign producers--a niche in which the domestic
producers had offered some products but which had been a minor
part of the business. The story of the small car in the American
market illustrates many of the themes developed in our discussion
of product convergence, international trade, and market
structure.[6] It is especially useful in providing insight into the
nature of the dilemma facing domestic manufacturers in 1980 and
the extent of the transformation that is under way in the industry.

Small Cars and Competition

The role of small cars in the U.S. market has been shaped by the
mode of competition that emerged in the 1920s and by the nature
of production technology in the industry. The struggle of U.S.
producers to produce a satisfactory low-priced, small, utility car--

one that would offer sufficient economies in material, design, or manufacturing to substain a significantly lower price than its full-sized cousin--dates back to the early 1900s.[7] In an early effort, Alanson Brush offered the 500 Brush runabout in 1907, featuring replaceable wooden axles and frames to reduce cost. Hard times in the 1930s brought a flurry of "downsizing" by Graham, Reo, Hupmobile, and others. In 1939, Powell Crosley, Jr., introduced his all new small car, the Crosley. Despite these efforts, the history of the small car in the United States does indeed support Detroit's generalizations of the 1950s--"small car, small profits." In fact, the graveyard of U.S. small-car companies would support even stronger condemnation of small-car economies. The reasons become clearer as the unique structure of the U.S. market is considered.

The fate of small cars in the U.S. market was strongly directed by the particular way that the market structure and production methods developed in the 1920s. The introduction of closed steel bodies in the 1920s changed the concept of the automobile and raised a whole new set of criteria for automotive design--passenger comfort, room, heating and ventilation, and smoothness and quietness of ride. As the car evolved toward the general-purpose road cruiser that dominated sales for 50 years, premium products were associated with larger size and greater weight. It was during this period that GM evolved its "full-line" product strategy and developed the organization, finance system, and pricing policies to support it. As quoted earlier, an important aspect of GM's strategy was its product-line pricing policy. In effect, market needs were to be met with adjustments in the entire product line rather than with independent design. More-over, serious effort was made to rationalize the coverage of price classes, to avoid overlap and market confusion.

When combined with the market's association of luxury and high prices with larger and heavier cars, the product-line policy resulted in a close and relatively smooth relationship between the price of a car and its size. Given GM's leadership and the pressure of competition, it is not surprising that this relationship pertained to the market as a whole. Indeed, as the automobile grew in size the relationship was, if anything, strengthened. This is illustrated in Figure 4.4, which presents a set of price-weight curves for the 1957-1961 period, the era of the first import wave.

The price-size relationship that grew out of market prefer-ences and product competition stands in sharp contrast to the relationship between size and cost. Within reasonable size ranges, production costs did not decline as rapidly as price with declining car size. The consequence was a growing price-cost gap in the small-size inexpensive car segment. Detroit recognized this as early as the late 1930s. The cost-size relationship is well

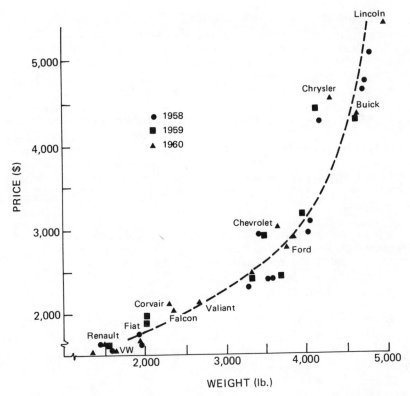

FIGURE 4.4 Price-weight relationships for selected models, 1958, 1959, 1960. (From <u>Consumer Reports</u>, April issues, 1958, 1959, 1960.)

illustrated in Ford's experience with the Model 92A in the late 1930s.

By 1936 no major producer was selling to the low-price market that had been served by the Model T. Ford undertook the development of the 92A to supply this market. Eugene Farkas of Ford's engine development group engineered the project. The car was to use the small, 136-cubic-inch displacement version of the V-8 engine first introduced to the market in 1937 and a scaled-down version of the Ford frame and body. The project was a technical success but an economic failure.

... Farkas engineered the model. He used the smaller V-8 engine, and the 92A, as the car was called, emerged narrower and shorter than the regular Ford, and 600 pounds lighter. The first completed model, as Farkas recalls, was a "sweet-running job." But difficulties arose. The small motor cost but $3.00 less to manufacture than the larger one. Wibel calculated the possible savings in each case at a mere $36. Since the 92A would have to compete with year-old larger used cars, this was not enough . . . so by mid April the project was abandoned.[8]

It appears that little of the cost of production depended on the size or weight of components; the only reduction in cost came from slightly lower material content. The relative insensitivity of cost to vehicle size is reflected in the testimony of L. L. Colbert, President of Chrysler in the late 1950s. Speaking before a Senate committee, Mr. Colbert responded to questions about small-car production:

> Up to this point all I can say is we at Chrysler have not given up, but we have not found a way yet to engineer, style, and build one of these smaller cars for enough difference in price to justify what we believe the American market demand for it is.[9]

It seems that without substantial redesign and new facilities geared specifically to small-car production, little reduction in cost could be expected from reducing the size of the vehicle.

Initial Import Penetration

In a market where size is associated with luxury and premium products and where pricing policy follows suit, the cost-size relationship renders small cars relatively unattractive from a profit standpoint. These relationships are illustrated in Figure 4.5, which contains a hypothetical cost-weight curve overlaid on the price-weight curve of Figure 4.4. The price-cost-weight relationship creates large profit margins on large cars and the possibility of losses on the very smallest models. The focus of the U.S. manufacturers on large vehicles is readily explicable in this context. Following World War II, managers at VW and Renault did not fail to understand that with favorable wage rates they could enter this vulnerable Achilles' heel of the U.S. auto market. So on the basis of very low factor cost and great attention to quality, VW launched a successful assault on the massive U.S. auto market. It was in light of such incentives that the domestic producers faced the first real penetration of foreign imports in the late 1950s.

The leader of the import surge was West Germany (largely VW).

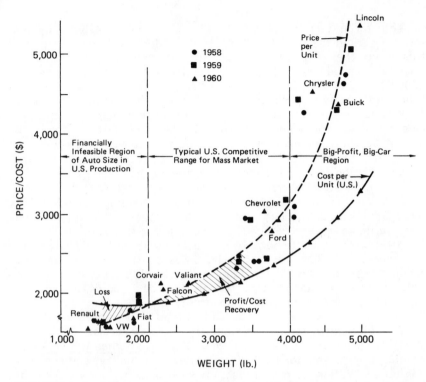

FIGURE 4.5 Price, cost, and vehicle size for selected models, 1958, 1959, and 1960. (From Consumer Reports, April issues, 1958, 1959, 1960.)

From a relatively small base of 38,000 vehicles in 1955, West German products expanded sales to 190,000 by 1960. With the additional sales volume accorded the French and British products, the share of imports reached 10 percent of the market in 1959. The imports were generally targeted at the low-price end of the market. The products were smaller, lighter, more economical in operation, and less expensive than the domestic product. The 1960 Ford Fairlane for example weighed 3775 lbs., and its lowest-price version sold for $2776. In contrast, VW had a 94.5-inch wheelbase, weighed 1650 lbs., and had a list price of $1595.[10]

The marked shift to small cars in the late 1950s benefited the smaller domestic producers, AMC and Studebaker, both of whom produced small cars--the Rambler and Lark, respectively. Rambler sales in 1959 were 363,000, while Lark sales totaled 133,000. Together with the imports the smaller cars accounted for 18.4 percent of the market.

The three largest producers all introduced newly designed smaller cars in the fall of 1959 (Corvair, Falcon, Valiant). Prior to this time they had participated in the small-car segment by importing products from their British (GM, Ford), French (Chrysler), and German (GM, Ford) subsidiaries. Although the new "compact" models were much smaller than the standard sedan, they were larger, heavier, and more expensive than the imports (see Table 4.6 for a comparison). The large domestic companies sought to fill a segment of the market just above the imports in terms of price and size, a strategy similar to that employed by GM in competition against the Model T in the 1920s.

The domestic compacts were quite successful. While some sales were taken from full-sized models, the imports were cut back significantly. A large part of the cutback occurred because captive imports were reduced to almost nothing. But other foreign manufacturers, most notably Renault, also suffered sizeable reductions in sales. Significantly, sales of the VW Beetle, the leading imported model, continued to grow. The success of VW in this era is particularly instructive in light of later developments. Although its low price was undoubtedly a key factor, two other aspects of the car were crucial. First, the "Bug" appears to have been well constructed, particularly in comparison with the domestic products. At a time (1960) when Consumer Reports was chiding the Corvair for its "unimpressive trim quality" and remarking on the Valiant's "poor finish," it was extolling VW's workmanship and construction. Second, the "Bug" was fun to drive. Again, Consumer Reports noted that the import's "handling and roadability are well ahead of the U.S. average."[11] Finally, VW had established a sales and service network that has been an important aspect of competition in the U.S. market.

Imports in the 1960s and the Entry of the Japanese

Before the success of the compacts was evident, Ford had begun development of a car to compete directly with the imports. Code named the Cardinal, the car was designed with a 96-inch wheelbase and, surprisingly, a front-mounted, four-cylinder engine with front-wheel drive.[12] The car was never produced in the United States but did emerge in West Germany as the Taunus 12M. At the time of scheduled introduction in 1962, imports had slipped to 5.0 percent of the market, and there was some evidence that consumers were "trading up" to larger, more luxurious versions of the compact models. Yet Ford's decision was unfortunate in the extreme, for the reduced demand for imports was to be temporary. The rise of the small car reflected fundamental demographic trends (increased suburbanization, shifts in the age structure, changes in female labor force participation) and the growth of multicar families.

The latent appeal of the imports was apparent in the growth of VW sales. The domestic compacts also grew, not only in sales volume but in size as well. The incentives inherent in the price-size connection as explained earlier were evident in subsequent operation. While in production the Falcon, for example, added several inches in overall length and several pounds in weight by 1968. By this time, however, imports, led by the Germans and the Japanese, had made a comeback. The second wave, which began in the mid-1960s, was much more soundly grounded than the first. Indeed, the strategy of low price, good quality, and a solid dealer network that had served VW well was refined and applied by the Japanese with obvious success. At the same time, Detroit was more vulnerable to foreign competition because of public disfavor brought on by politicization of the safety and emissions issues. Many U.S. citizens were eager to buy foreign as a protest.

The creeping growth of the compacts left the domestic producers without products to compete with the imports. Before new products were introduced in the early 1970s, imports took 15 percent of the market. With the introduction of the Vega (GM), the Pinto (Ford), and the Gremlin (AMC), domestic manufacturers began to compete directly with imported products. Yet even in the early 1970s the U.S. cars were heavier and had larger engines. Unlike Ford in 1936 the domestic manufacturers did not simply miniaturize the larger cars. The new domestic subcompacts were fundamentally redesigned to reduce the number of parts by 30 percent and were produced with higher levels of automation and capital intensity than was typical of industry practice. These changes reflected the need to shift the cost-size curve down in order to improve the prospects for profitable production of small cars.

Small Cars and the Crisis of 1980

The history of the small car in the U.S. market prior to 1973 underscores the extent of the transformation under way in the domestic industry and illustrates well the way in which pressure for internationalization and convergence in products has affected the U.S. market. Given the character of the market and the technology that prevailed for most of the industry's history, small-car production was both unprofitable and difficult to incorporate successfully into the product strategies of the domestic manufacturers. It is clear in retrospect that Detroit has always had the capability to produce a small car, but its dominant orientation both in production and in marketing--at least in domestic operations--has been elsewhere. Thus, to successfully produce a small car would seem to have required some basic changes--partly

TABLE 4.6 A Comparison of Imported and Domestic Products in 1960

Model	List Price (dollars)	Weight (pounds)	Wheel-base (inches)	Width (inches)	Front (inches)	Rear (inches)	Engine Displacement (cubic inches)	Consumer Reports' Comments
Domestic Midsize								
Chevrolet Bel Air	2893	3690	119	81	65.5	65.5	283	A smooth out-size package with good-quality trim and freedom from new-model bugs.
Ford Fairlane	2766	3775	113.5	81.5	62	63.5	292	Serviceable vehicle, although mediocre in most characteristics.
Plymouth Belvedere	2899	3860	109.5	78.5	63	62.5	318	A good, fast road-car choice.
Domestic Compacts								
Chevrolet Corvair	2112	2325	108	67	58	57.5	140	A personal car, a fun car; body dimensions—low seats, low weight; rule out use as a family sedan; add Corvair's unimpressive trim quality you have reasons for low rating.
Ford Falcon	2042	2355	109.5	70	57	57	144	Smartly conceived small-compact package with common-sense design; steering and transmission not good.

								Comments
AMC Rambler	2131	2650	100	73	58	45	196	Top-quality, top-compact size, seasoned design; repair record outstanding; 25 MPG at 50 mph.
Plymouth Valiant	2127	2680	106.5	70.5	57	57	170	High marks as a well-rounded behaved design, but haunted by poor finish and hasty assembly.
Imports Volkswagen	1565	1650	95	61	N/A	N/A	73	Unbeatable quality for price, good workmanship, unexcelled durability, long-legged speed. Service widespread, resale value abnormally good; 32 MPG at 60 mph.
Renault	1645	1500	89	60	N/A	N/A	52	Interior finish serviceable rather than attractive or of high quality; excellent road manner, good ride and operating economy; 37.8 MPG at 60 mph.
Fiat	1659	1965	92	57	N/A	N/A	67	Car offers ruggedness, solid workmanship and adequate but not outstanding behavior; a very compact, respectworthy small car well finished and well equipped; 28.8 MPG at 60 mph.

SOURCE: *Consumer Reports*, April 1980.

in technology, partly in organization and strategy. Perhaps most important, it required a restructuring of the market-pricing structure so that small cars could be sold at a profit. This might have been difficult in earlier years but ultimately had to be faced. This is clear in marketing, where an organization long oriented toward "trading up" in size, associating premium with size, and pricing products across a range of size classes must sell and sell profitably in a market where variations in size are sharply diminished, where everything is small.

In retrospect, it is evident that the oil shocks of the 1970s have brought the U.S. market for automobiles into the international arena. Differences in tastes and relative prices that insulated the domestic producers from extensive import competition have changed dramatically. This internationalization has confronted the U.S. firms with many more competitors, and it has required major changes in operations and strategy. This is true not only because the market has changed but also because the new competitors approach competition very differently. The old days, where the three main producers competed on a relatively small number of dimensions (e.g., styling, dealership performance, economies of scale) in a market focused on the large road cruiser, are gone. The industry is now faced with several somewhat unfamiliar competitors with different strategies and a market demanding different products.

Much has been made of the lack of U.S. capacity for producing the types of products in demand in 1980 and the large investments required to obtain it. Yet it is also useful to recognize that competitiveness in the future is likely to require changes in orientation and organization that may rival the billions in capital investments in importance. In some ways the challenge facing the industry is similar to the challenge faced by producers in the 1920s, when changes in the product and the market transformed the industry. Success in that era demanded not only new products but also new forms of organization and a new strategic posture; survival in the modern era may demand no less.

NOTES

1. See Wilkins (1980) for a review of the effect of government regulation on international trade.
2. These arguments have been developed in unpublished work by David Haddock.
3. Abernathy (1978), pp. 29-31, notes the trends in concentration from the early days of the industry.
4. See Vernon (1977) for data on equivalent firms for a number of worldwide industries.

5. See Fuller (1981) for a discussion of these relationships.

6. White, L. (1971) presents an analysis of the small-car story that emphasizes the effects of market structure on new product development.

7. See Flink (1970) for information on early automobiles and attempts to develop a small car.

8. Nevins and Hill (1962), pp. 117-119, cited in Abernathy (1978), p. 28.

9. Testimony by L. L. Colbert, cited in L. White (1971), p. 184.

10. Consumer Reports, April 1960.

11. Ibid.

12. The Cardinal program is mentioned in Tracy (1978b).

5

Government Regulation: The Evolution of Public Demands on the Industry

THE ORIGINS OF REGULATION

The development of automotive technology during the last 10 years has been strongly affected by government mandate. A variety of congressional committees and regulatory agencies have issued rules and standards intended to enhance safety, reduce air pollution, and cut fuel consumption. Although new in kind and degree, these demands on the auto industry are hardly without precedent. Almost from its inception, the automobile has had a far-reaching influence on the life of the nation. As a result, manufacturers have long had to meet both the evolving demands of the marketplace and the requirement of changing social expectation, whether expressed in the form of explicit government action or merely of diffuse public sentiment.

Public Demands in the Formative Years

At the turn of the century, many hailed the automobile as the guarantor of public health in urban areas and awaited the arrival of the "horseless age" with high expectations.[1] Yet the American romance with the automobile sprang from deeper motives than the desire of city dwellers to be rid of the horse. First and foremost, the automobile statisfied a pervasive desire for personal mobility. Horses and bicycles had obvious limitations; trolleys and railroads were rigid and inflexible. Moreover, rail-based transportation appeared to the public not only as monopolistic, corrupt, and unscrupulous but also--given its insatiable need for rails, tunnels, and overhead wires--as capital intensive, cumbersome, and centralized. In contrast, the automobile was quick, inexpensive, and immensely flexible. Timing and destination were at the discretion of the individual. A transportation system based on the car was at once democratic and decentralized. It required roads but little else.

Although the desire for personal mobility was the principal force behind private demand for automobiles, that demand had a social dimension as well. In an age of growing urbanization and industrialization, the car seemed a solution to many of the problems of large cities. It removed the horse, gave people access to the countryside, and offered (in the words of one proponent) to "eliminate . . . the nervousness, distraction and strain of modern metropolitan life."[2] In turn, suburban living could become a quality, and some observers waxed eloquent in their descriptions of it:

> Imagine a healthier race of workingmen, toiling in cheerful and sanitary factories, with mechanical skill and trade-craft developed to the highest, as the machinery grows more delicate and perfect, who, in the late afternoon, glide away in their own comfortable vehicles to their little farms or houses in the country or by the sea twenty or thirty miles distant! They will be healthier, happier, more intelligent and self-respecting citizens because of the chance to live among the meadows and flowers of the country instead of in crowded city streets.[3]

In a heavily urban, industrial society, still wedded to the values of rural/agrarian life, the automobile had a social value beyond its private appeal. Between 1910 and 1925 the widespread use of the car was viewed as a progressive force in America, and its manufacturers were accorded the respect due members of an industry that effectively met at a reasonable price the demands made of it.

The private and public demands facing the industry in its early years were consistent and mutually reinforcing. Antitrust activity, no less than the desires of the consuming public, favored the mass production of automobiles. In 1903, Henry B. Joy of Packard, Frederick Smith of the Olds Motor Works, and several other prominent manufacturers joined with the Electrical Vehicle Company, which held the Selden patent on the gasoline automobile, to form the Association of Licensed Automobile Manufacturers (ALAM).[4] ALAM members, largely high-priced producers catering to the luxury market, sought to limit entry by granting licenses only to manufacturers with prior experience in the business. Although a number of independent producers disregarded the Selden patent and entered the industry, the ALAM held 80 percent of the market in 1907.

Early in its history, ALAM rejected Henry Ford's application for a license and thereby earned itself an implicable foe. Ford, which sought to produce a car for the great multitudes, fought the patent and the ALAM almost singlehandedly. Court action initiated by ALAM led to a decision in Ford's favor in the federal courts that hastened the demise of ALAM. When Ford himself came to dominate the market with his Model T, the integration of production proved a far more effective barrier to entry than the Selden patent, and there was no hint of antitrust activity. Both public policy and private need smiled on the growth of that kind of large-scale enterprise.

A similar convergence of interest emerged with the federal government's long-term commitment to building and upgrading roads to complement the use of the automobile. The sheer magnitude of the expenditures was extraordinary, but more remarkable still was the widespread popularity of state and local taxes--on property as well as gasoline--to support road construction. As Flink has remarked:

> Public support for heavy motor vehicle special use taxes is a case in point. Motorists early came to support higher and annual registration fees as one means of securing better roads. For the same reason, there has consistently been almost no public opposition to the gasoline tax. By 1929 all states collected gasoline taxes, which amounted to some $431 million in revenue that year, and rates of three and four cents a gallon were common. In 1921 road construction and maintenance were financed mainly by property taxes and general funds, with only about 25 percent of the money for roads coming from automobile registration fees.[5]

The New Deal and the War Years

Public demands on the industry in the New Deal era were little changed from the 1910-1925 period. Expanded highway construction further enhanced personal mobility, and the rhetoric of some New Dealers (even FDR himself) continued to idealize the notion of new communities that would combine the best of rural and urban life. Although the bulk of New Deal legislation (e.g., the National Recovery Act) had only minor effects on the auto industry, the passage of the Wagner Act, which legitimized collective bargaining, led to the organization of unions at General Motors (GM) and Chrysler in 1937 and Ford in 1941. Unionization has, of course, been of obvious importance in the industry's development, but in terms of products developed, markets served, and strategies employed, the social demands made on the industry during the 1930s largely continued trends begun much earlier.

With the coming of World War II the industry's public responsibility was direct and clear cut--mobilize. The conversion to war production involved not only changes at the plant level but also the transfer of many top executives to high-level government positions. For example, William S. Knudsen, President of GM, was put in charge of war production in the War Department. And the industry produced. The magnitude of its contribution was striking, millions of guns, trucks, tanks, engines, and airplanes--in all, over one-fifth of all defense production. As John B. Rae has argued:

> The automobile industry was the country's greatest reservoir of "know how" and skill in the technique of making, accurately and reliably, the largest possible number of items in the shortest time.[6]

A Shift in Perceptions

At the end of the war, most agreed with Rae that U.S. superiority in mass production techniques had been a major factor in the successful war effort. The auto industry was viewed as a valued national resource; its leaders were called on to serve in responsible public positions; its capabilities were admired and respected.

This attitude lasted into the 1950s. In the period following the Korean War, a variety of concerns about the automobile industry and its impact on society began to surface. Initially focused on dealer practices, public scrutiny of the industry shifted to issues of pollution and safety.[7] The emergence of these issues reflected, in part, the maturity of the industry. Not only were consumers becoming more sophisticated, but the sheer size of the U.S. car fleet made the side effects of driving more noticeable. Thus, while fatalities per mile driven were either stable or falling in the 1950s, the total number of fatalities increased by 50 percent, reaching 30,000 by 1956. Likewise, in large urban areas, most particularly Los Angeles, the deterioration of air quality was noticeable.

At the outset, pollution and safety were not burning national issues. Newspaper coverage was infrequent and often buried in the back pages; congressional involvement was limited to modest authorizations for studies of the health risks imposed by smog and to a series of hearings before a small House subcommittee on traffic safety. Yet even these modest probes of the industry and its product marked an important change in the social demands made on the manufacturers. The mid-1950s witnessed the beginning of divergence between private desires, as expressed in the market, and public concerns. At a time when the market sought larger, more powerful, more exciting automobiles, society generally began to question the effect of such cars on public health and safety.

Disagreement and debate were perhaps inevitable, yet there was nothing inevitable about the form of that debate or of its results. Somewhere between 1953 and 1970 the public view of the industry was transformed. Its image of dynamic growth, superior technology, and progress gave way to one of unprincipled social irresponsibility.

The Emergence of Regulation by Mandate

The public search for cleaner air and safer highways emerged in full force in the mid-1960s. A burgeoning environmental movement, a growing aversion to large institutions and concentrations of power, and a backlash against wealth and conspicuous consumption made the automobile an easy target. Political points could easily be scored by attacking the industry on its safety record and on pollution, and this politicization of the issues had an enormous impact.

The issue of safety is instructive.[8] Rising numbers of auto fatalities in the early 1960s brought the issue of auto safety under greater public scrutiny. Hearings on auto safety were initiated in the Senate during 1965, primarily under the auspices of Senator Ribicoff's Subcommittee on Government Operations. Industry representatives reported on their companies' efforts to increase auto safety and stressed the need to include the effects of roads and drivers in any consideration of traffic safety. In November 1965, Ralph Nader's book, Unsafe at Any Speed, indicted the industry for what Nader believed was a callous disregard of the consumers' "body rights."[9] The book helped focus government attention on the role of vehicle design in crash survivability.

In early 1966 President Johnson called for a highway safety act to "arrest the destruction of life and property on our highways." Senator Ribicoff again held hearings on safety, and the Senate Commerce Committee heard testimony on the administration's bill. Those hearings were conducted in a heavily politicized atmosphere. A few days before they began, GM's investigation of Ralph Nader was revealed, and GM executives were summoned before Senator Ribicoff's committee. Their lame explanations helped fix in the public mind the image of a big corporation harassing a concerned citizen. Before the GM-Nader incident the passage of some sort of legislation mandating regulatory standards had by no means been certain. Now it was.

The administration's bill called for the Secretary of Commerce to set federal standards for equipment if, after two years, he determined that the automobile industry had not developed adequate standards of its own. As the hearings before the Commerce Committee got under way, the auto industry endorsed the goals of the administration's bill but suggested, instead, a voluntary plan. The political climate would not allow this, and the industry shortly changed its position in testimony before the House Interstate and Foreign Commerce Committee and supported federal authority to set safety standards. The Senate then passed a bill calling for mandatory standards, and President Johnson signed it.

Partly because of GM's response to Nader and partly because of the political climate, the public demand for safer vehicles came to be embodied in a regulatory process involving mandatory standards, with the government and industry essentially in opposition. The adversarial nature of the process was further sharpened in the debate over pollution. The government's first effort to control emissions, the 1965 amendments to the Clean Air Act of 1963, gave standard-setting authority to the U.S. Department of Health, Education, and Welfare (HEW). The initial legislation gave due weight to economic and technical considerations, and the regulations eventually developed in 1966 set fairly long-term standards that the industry believed it could meet. Industry optimism was short lived. In January 1969 the Justice Department charged the major producers with conspiracy to delay development of devices to control pollution. Settled by a consent decree in September 1969, the suit tarnished the industry's image and changed its relationship with the government.

In November 1970 the Senate, taking stock of new realities, passed amendments to the Clean Air Act that established a standard of 90 percent reduction in pollutants over allowable 1970 levels to take place by 1975-1976. Furthermore, the Senate required that control devices be effective for 5 years or 50,000 miles and that administration of the law be taken from HEW and placed in the hands of the newly created U.S. Environmental Protection Agency (EPA).

The law's provision for an optional delay of the standard by EPA led to a long series of public arguments, requests for extensions, and judicial and administrative proceedings, which resulted in a one-year extension. In the course of this debate, industry representatives made a series of arguments that subtly reinforced the image of footdragging and reluctance. The industry position progressed from "technologically it can't be done" through "the technology is untried and untested" to "it can be done, but it will cost so much it is not justified." Elliot Estes, President of GM, summed up the effects of the industry's approach:

In dealing with the government--and in raising questions and explaining the possible difficulties and costs, we have reinforced the negative image that many people have of us--I don't know how it can be avoided.

In all honesty, we have contributed to this lack of credibility because we wanted to see some promising results with real hardware before we predicted our ability to make progress in meeting some of these standards and rules.[10]

The pattern of regulation established for emissions has strongly influenced the government's approach and the industry's response to fuel economy. Choosing not to rely on taxes or the price of fuel to spur demand for smaller, more efficient vehicles, the government opted for direct regulation of fuel economy through the setting of standards by Congress and the administration of those standards by an executive agency.

The decisive year was 1975. After several months of public statements, hearings, and proposals, President Ford obtained in early 1975 voluntary commitments from the major producers for a 40 percent fuel economy improvement by 1980 in exchange for a five-year moratorium on emissions standards. The industry hailed the agreements, but Congress proceeded to advance more stringent requirements. Efforts to impose various kinds of taxes on less efficient cars were discarded in favor of a bill requiring a mandatory level of average fuel economy for the corporate fleet. The industry questioned the viability of imposing cars with specific characteristics on a market that might not want them. Mandatory standards grew in political appeal, however, and President Ford abandoned his earlier agreements and signed the Energy Policy and Conservation Act of 1975 into law.

An Adversarial Environment

The motor vehicle regulatory environment that emerged in the 1970s is best characterized as a combination of congressional and agency rulemaking with administrative and judicial review.[11] It is an inherently adversarial process, one that relies on the ex parte use of political power to achieve social objectives, for it took shape in an era when public opinion viewed the industry as a "bad guy" that had to be closely regulated. Indeed, the legislative record suggests that some members of Congress and their staffs have typically operated on the assumption that, if the industry does not oppose it, it must be too lenient.

The automobile companies must, of course, bear partial

responsibility for this poor relationship with government, but the explanation cannot simply be bad judgement, irresponsible behavior, or a lack of moral fiber in their leaders. There is, first, the growing divergence in the 1953-1975 period between the private and public demands made on the industry. A good part of the industry's position on safety emissions and fuel economy sprang, after all, from its desire to meet perceived market demands. But there is also the substantial ambiguity inherent in the regulatory process itself.

The regulation of emissions, safety, and fuel economy involves ultimate objectives (e.g., safe highways) that are relatively uncontroversial, but practical means (e.g., specific equipment standards) that are open to debate.[12] Further, even where standards and objectives are clearly linked (e.g., Corporate Average Fuel Economy standards), there may be technological uncertainty associated with both production and performance.[13] Recent regulatory history has been filled with technological "optimism" on the part of regulators and "pessimism" on the part of the manufacturers. Experience has shown, however, that neither position is always justified. The development of the catalytic converter allowed the auto companies to meet emissions standards that they had once claimed were impossible, yet new fuel-economy goals created technical difficulties in meeting emissions targets by established deadlines.

THE IMPACT OF REGULATION ON COMPETITION AND INNOVATION

Government involvement in safety, pollution, and fuel-economy decisions played a significant role in the design and manufacture of automobiles in the United States in the 1970s. A full analysis of the impact of regulation--on objectives and on overall economic and social welfare--is far beyond the scope of this report. It does seem clear, however, that the rules and laws adopted have not been neutral in their impact on competition or on the introduction of new technology.

Competition and Regulation

The form of regulation governing the automobile in the United States--mandatory standards administered by an executive agency--imposes a single set of standards on companies employing different competitive strategies and enjoying quite different capabilities.[14] Such regulation inevitably affects each firm differently and thus alters relative competitive positions. A brief discussion of the difference between the GM and the Chrysler positions on the 1970 amendments to the Clean Air Act highlights the differential strategic impact of regulation.

From GM's perspective, the catalytic converter--and the legislation to make it mandatory--had several advantages. It was a familiar technology and, as an add-on device, limited the need for fundamental changes in established manufacturing and assembly skills. Furthermore, as a backward-integrated firm, GM could view catalytic converters as a source of profit. Converter technology reinforced existing GM strategic strengths and did not make existing corporate strategy obsolete.

Chrysler had a different strategic exposure to catalytic converters. The technology was relatively unfamiliar; as an add-on device it did little to create an opportunity for Chrysler to improve its competitive position against other domestic producers; as one of the least vertically integrated domestic producers, Chrysler had little opportunity to capture any value-added in manufacture or assembly of the devices.

Chrysler's preferred compliance technology, which involved electronic technology and the lean-burn engine, played to its historic strength as an engineering-oriented firm. It did not penalize the firm for its lack of vertical integration and, indeed, made the relatively inflexible backward-integration strategies of its competitors less attractive. As a knowledgeable innovator in this area, Chrysler looked on such a technology as a competitive opportunity, just as GM did the catalyst. Government standards, however, finally ruled out all options but the catalytic converter. This kind of competitive consequence is a general feature of regulation in an environment where firms pursue different strategies and possess different kinds of technical competence. While proponents of regulations can (and do) claim some measure of success in forcing the adoption of the catalytic converter (among other innovations), the impact on competition and the relative success of the producers must also be weighed in the balance.

This becomes particularly evident in comparisons of domestic and foreign producers. When regulation is introduced as a factor in international competition, it is often said that since all competitors must meet the same standards, regulation must be competitively neutral. However, where stationary regulations (e.g., those of the U.S. Occupational Safety and Health Administration, and EPA's water and air pollution regulations) are not comparable, the overall regulatory impact may be different in different countries. Moreover, the notion that regulation is neutral ignores the fact that firm capabilities and circumstances are not identical. In a period of crisis and transition, such as the current one, heavy demands are placed on scarce resources simply to survive. Regulation that competes for those resources but that does not enhance a firm's competitive position becomes an added drawback. Although much has been made of the capital cost of regulation, the more critical and scarce resource is likely to be the time, energy, and attention of knowledgeable and talented individuals.

Innovation and Regulation

Proponents of government involvement in product design often point to a series of innovations that have emerged in response to mandated standards. And it is true that a number of technical advances can trace their origins (at least in part) to regulatory action. Yet in a more general sense it is not clear that the pace and character of innovation is necessarily enhanced by mandated standards.[15] To see the potential barriers to change inherent in regulation, it is important to distinguish between the radical or epochal innovations characteristic of the early stages of production development and the incremental innovations that dominate as a product matures. These two patterns of development may be observed at the same time when radically new products are introduced that challenge existing technology.

Epochal innovation involves the identification of new needs or a new way of meeting old needs; it is essentially entrepreneurial in nature. It competes with the existing technology on the basis of performance rather than cost. Because markets for the new product are apt to be ill defined and because the manufacturing process is apt to be both labor intensive and fluid, the entrepreneurial firm may continue to make dramatic changes in the new concept. In this context, thin specialty markets play an important role in the development and commercialization of a new technology. Buyers in such markets share common traits: (1) a willingness to pay high premiums for superior performance in a few limited dimensions and (2) a willingness to accommodate some performance deficiencies in the new technology compared with existing competitors.

It is important to understand how the relationship between thin, performance-oriented markets and established mass markets affects the process of successful innovation and commercialization. At the point of introduction the new product is very vulnerable. It is often introduced by small, entrepreneurial firms or organizations that lack the resources to undertake major risks or to sustain high rates of R&D expenditures. The greater the established product's economies of scale and production volume, the greater the need for robust specialty markets to nurture innovations until they are able to compete within established markets.

Government regulation can alter the innovation process through its effect on thin specialty markets. While the inhibiting effects of regulation can be subtle, they are nonetheless powerful and pervasive. The normal patterns of interaction between thin, high-performance markets and established markets may be disrupted in three distinct but related ways: (1) barriers to the initial development of new competing technology may increase, (2) existing technology may be enveloped with regulatory requirements so that no new technology can fully satisfy the web of constraints so created, and (3) regulation may encourage the entrenchment of current technology within the industry by diverting all discretionary resources to improve existing technology.

The most obvious and frequently cited consequence of regulation on the innovation process is the barrier erected to the initial development of new products. This barrier results from the increase in resources, costs, and risks involved in developing and introducing new technologies. Turbochargers in California provide a useful example.[16] In the 1960s California erected a regional barrier to turbocharger development that had national implications. According to market surveys, California was the largest potential after-market for turbochargers, but the state prohibited turbocharger installation pending certification by its Air Resource Board. Certification required a durability test of at least 30,000 miles and thereby imposed requirements that were simply too complicated and costly for the small firms manufacturing turbochargers. Thus, as a direct consequence of regulation, the thin speciality market that California offered for developing turbochargers for automotive passenger cars never materialized, setting turbocharger development back a number of years.

Experience with air bags and air brakes suggests that uncertainty over standards also can serve as a barrier to innovation by supplying firms. As previously noted, the problem of setting clear and certain requirements is inherent in the nature of the regulatory process.

Though regulation may increase barriers to new firms seeking entry into the industry with innovative technology, it may also affect the involvement of established firms in the innovation process. Steadily tightening regulatory requirements forces companies to divert discretionary resources into programs to improve existing technologies, in effect entrenching the current technology within the industry. While this encourages more rapid incremental innovation, it may also discourage the entry of firms undertaking needed longer-term advances or epochal innovations. An intensification of regulation, whether by adding new kinds of requirements or by tightening existing ones, requires the manu-facturer to devote even greater resources to the existing technology and market. As new requirements create new demands, R&D tasks associated with each change become more complex, costly, and subject to risks. Each change, too, becomes more costly while at the same time more changes are required.

This escalation of development cost and complexity is clearly evident in the engineering interactions on new engine development: new requirements and components interact with each other so that the effect on the number of subordinate design tasks, tests, and, ultimately, costs is more nearly multiplicative than additive. For example, the interaction of tough fuel-economy and emissions requirements for automotive engines has led to the addition of much more complex engine-control technology and carburetion systems, as well as catalytic converters and related components. Similar effects are reported for other drive-train components.

The causes of entrenchment are subtle; their consequences, however, are vitally significant for firms in the industry. For example, to obtain the resources it needs to compete successfully in the highly regulated U.S. market, Chrysler has divested itself of many of its extensive foreign operations. In Ford's recent report on the state of the automobile industry, it documents a need for an additional billion dollars (adjusted for inflation) above its recent, historically high rate of capital investment in order to remain competitive in North America through 1985.

By far the most subtle influence of regulation on the innovation process is regulatory envelopment. The stream of automotive regulations in the last decade has broadened substantially from the minimum safety and pollution-control regulations of the early 1960s to the more extensive standards and rules of the late 1970s; Eugene Goodson has counted 237 regulatory changes pertaining to automobiles and light trucks from 1960 through 1975.[17] In this evolution of standards and rules, regulators have often favored performance regulations over design standards in order to preserve the manufacturers' freedom to innovate. They have also limited regulations to specific objectives and based them on the best available technology. In attempting to protect the innovative process by undertaking piecemeal regulations, however, government agencies may have achieved the opposite result. They may have created a sequence of independent regulatory actions that, taken as a whole, form a tightening web of constraints that envelop the existing technology.

Fragmented performance regulations issued by different organizations become an overall design standard when the automobile is considered as a single, integrated system. This unified design standard bars the entry of initially imperfect but potentially useful new technologies. The barrier effect may thwart the initial development of a new technology; envelopment bars the acceptance in established markets of such innovations as are made.

Honda's CVCC program illustrates how envelopment may lead even a highly creative company to innovate incrementally.[18] In their search for an engine concept that offered a competitive edge under impending U.S. and Japanese fuel-economy regulations, Honda's engineers rejected more radical engines such as the Wankel, steam, and electric. These engines were incompatible with the emissions, durability, cost, produceability, and fuel-economy profiles of current engines. Honda's engineers decided to develop instead the 50-year-old idea of charge stratification, relying on a particular combustion chamber configuration much like the Russian production-model Nilov engine. This more incremental approach has been successful.

The case of electric vehicle certification under Section 212 of the Clean Air Act provides a contrasting example of blocked innovation. In an effort to encourage innovation with respect to emissions requirements, Congress authorized the U.S. General Services Administration (GSA) to pay a premium of more than 100 percent for low-emission vehicles to be used by federal agencies. In effect, Congress attempted to create through federal procurement policy a thin specialty market. Only three manufacturers of low-emission vehicles applied. All three offered electric vehicles that certainly met the emission requirement but that failed to meet other regulatory and GSA performance standards, which were based on vehicles then in use by the government. None of the applications led to Section 212 purchases. It is important to see the contradiction at work here. While the legislation provided a price incentive to support an essential but thin, performance-oriented market, it neglected to protect the developing product by relaxing regulations or standards geared toward the existing technology.

The existence of thin, high-performance markets has been of importance in the process of innovation. In light of the role of thin markets in furthering radical innovation, regulation often creates a paradox: while encouraging more rapid progress through incremental innovation in established products, intense regulatory pressure can also inhibit epochal innovation by its effect on thin markets, by increasing barriers to development of new technologies, by entrenchment, and by enveloping existing technologies.

NOTES

1. This section draws on the work of Flink. For a more extensive discussion, see Flink (1970) and (1975) and the sources he cites.

2. Flink (1975), p. 39.

3. Ibid., p. 40.

4. A thorough treatment of the ALAM is found in Rae (1959), pp. 67-85.

5. Flink (1975), pp. 149-150.

6. Rae (1965), pp. 152-159.

7. For a discussion of the politics of safety regulation, see Halpern (1972).

8. The chronology used here follows that developed in unpublished work by Karen Tracy of the Baker Library, Harvard Business School. For emissions and fuel economy, see Tracy (1978). Additional insight is provided in several articles included in Ginsburg and Abernathy (1980); see especially those by Mills, Seiffert, and Kaspar.

9. Nader (1965).

10. Cited in Tracy (1976).

11. Implications of this form of regulation for fuel economy are examined in John et al. (1980), pp. 118-143.

12. Disagreement over process is examined in Leone et al. (1980) and in Mills (1980).

13. See Leone et al. (1980).

14. This section draws on the analysis presented in Leone et al. (1980); see also Hanson (1980).

15. The subject of innovation and regulation has generated substantial literature. The perspective presented here relies on concepts and analysis developed by Abernathy (1980); this section condenses one argument presented by Abernathy.

16. See Ronan and Abernathy (1978a) for an extensive treatment of the introduction of the turbocharger.

17. Goodson (1977).

18. Ronan and Abernathy (1978b).

6
The Sources of Competitive Advantage: Cost and Quality Comparisons

Almost 30 percent of all new cars sold in the United States in the spring and summer of 1980 were manufactured outside North America. The recent surge in imported products was influenced by a series of special factors that caused a rapid shift in consumer preferences. Yet the level of import penetration has been growing since the late 1960s. It seems that the long-term success of the Japanese and European producers may be the result of more fundamental factors than the recent shift in preferences or the absence of domestic small-car capacity. We noted in Chapter 4 that domestic producers have had little incentive to excel in small-car production and that at least until the early 1970s the foreign producers typically enjoyed a cost advantage. The evidence also suggests that the most successful foreign producers [e.g., Volkswagen (VW), Toyota, Nissan] have combined lower costs with an emphasis on quality. Thus, initial penetration was achieved at the low-price end of the market, but the imports sought to add more performance and quality to the product than the low price itself might have suggested. Rather than pass their cost advantage on to the consumer in the form of an even lower price for an average-quality product, the imports used part of their cost advantage to develop a more competitively viable advantage in product performance and quality.

The success of the import strategy provides useful insight into the nature of competition and consumer demand in the market for smaller cars. (The analysis may also apply to other segments, but our focus is on the markets in which domestic and foreign products compete.) The trend in import penetration suggests that the market is sensitive to the price-quality package and that product quality is becoming an increasingly important dimension of competition. It also seems clear that production cost and product quality are closely related and that both must be examined in assessing competitive advantage. Comparison of the relative competitive position of domestic and foreign products is in essence a comparison of systems of production. The success of the imports has underscored the obvious point that a production system must be judged in terms of the cost and the quality of its output.

The twin issues of cost and quality are central to the future of the domestic auto industry. An assessment of the continued viability of domestic production requires an evaluation of current competitive positions and some analysis of trends in the underlying determinants. We first consider the costs of production. Our approach is to draw on publicly available information as well as on data from industry sources to arrive at estimates of the cost of producing a comparable vehicle in both the United States and Japan. We then examine evidence on product quality from a variety of sources. Several dimensions of quality are identified, and an attempt is made to assess the relative position of domestic and foreign products on each. The chapter concludes with an assessment of the sources of U.S.-Japanese differences.

It should be noted that we do not attempt to assess the comparative advantage of U.S. producers in the sense of classical economic trade theory, which depends on relative costs of production at home and abroad (e.g., ratio of costs in autos to costs in other goods, versus the same ratio in other countries). Our analysis is focused on the competitive position of U.S. automobile producers relative to their major competitors within the automobile industry. The approach we use here places priority on understanding the characteristics of intraindustry competition, including the role of costs, product quality, and technological innovations. While an analysis of domestic relative costs (i.e., autos versus others in the United States) and relative costs in Japan or Europe would be a potentially useful element in an assessment of trade patterns, such an analysis in the context of differentiated products, economies of scale, and oligopolistic markets is likely to be complicated and is beyond the scope of this study.[1]

COMPARATIVE COSTS OF PRODUCTION

Over the last several years, information on foreign and domestic productivity and factor prices has been developed that implies a slight advantage for Japanese products and a disadvantage for producers in West Germany and the United Kingdom. These comparisons are for vehicles available for sale in the United States and thus include the costs of ocean freight and applicable tariffs for imported products.

One of the most careful studies of relative costs was conducted by Eric Toder and his colleagues at Charles River Associates.[2] Using data for 1974, Toder found a Japanese disadvantage of 3 percent. If Toder's analysis were revised to reflect more realistic transportation costs, the data would imply a Japanese advantage of about 7 percent.

In 1978, additional U.S.-Japanese cost comparisons were published by Ford in a white paper entitled State of the U.S. Automotive Industry. Ford estimated the net Japanese cost advantage per vehicle to be $525 on a subcompact-size car landed in the United States. Although methods and sources were not identified, higher U.S. costs were largely due to higher wage rates, a result generally consistent with the analysis of Toder. The difference in the two analyses could be explained by differences in time period, intervening inflation, and changes in productivity.

The notion that the landed-cost differential between U.S. and Japanese products is $500-$600 recently found its way into congressional testimony. Speaking before the Subcommittee on Trade of the House Ways and Means Committee, Abraham Katz, Assistant Secretary of Commerce for International Economic Policy, summarized what appears to be a concensus view:

> Average hourly compensation (including fringe benefits) in the Japanese auto industry in 1979 was $6.85--half of the $13.72 hourly compensation in the U.S. auto industry. Present indications are that productivity in the U.S. and Japanese auto industries may be roughly equal. On this basis Japanese producers appear to have had an $860 labor cost advantage per car in 1979. Other differential costs (principally the higher cost of steel in the United States) may have added $100 per car to the U.S. cost. As freight and insurance on Japanese cars averages $400, the apparent cost advantage to Japanese producers may have been $560 per car in 1979. The actual advantage may have been considerably less, for the above calculations do not take into account energy costs, capital costs, and the costs of other production factors--some of which are cheaper in the United States than in Japan.[3]

It is our view that the estimates presented by Katz and his suggestion that actual differences might be even lower constitute an understatement of the current cost advantage of the Japanese. Not only do the estimates fail to reflect current rates of compensation but they also fail to capture important differences in production processes that result in higher productivity in Japan. Estimates that reflect these differences have been developed using a variety of methods. We have estimated the productivity and cost differential using both a macro, economy-wide approach and a "bottom up" approach with microdata. We have taken an industry-wide perspective, using publicly available sources, and we have analyzed annual reports. Comparisons of this sort involve several difficulties. The automobile manufacturers of the United States and Japan produce a different mix of products and have organized production in different ways, particularly in terms of vertical integration. Productivity comparisons are also significantly affected by differences in capacity utilization that have been substantial in recent years. While attempts have been made to correct for these factors, even the most careful comparison requires judgements and assumptions that affect the results. A detailed description of the various kinds of analysis used is presented in Appendix A.

Evidence from Alternative Perspectives

Table 6.1 outlines the various perspectives taken and summarizes the basic results. The analysis suggests that the Japanese enjoy a landed-cost advantage of between $700 and $1500 per small vehicle. These estimates are larger than those used in congressional hearings during 1980. Furthermore, the immediate sources of the Japanese advantage may be quite different. Those analyses that focused on the auto sector found sizeable differences in labor hours per vehicle (a measure of productivity) along with differences in employee costs and other prices. The industry level analysis (which includes suppliers) found a 20-25 percent Japanese advantage in productivity, while examination of specific plants and processes revealed an even larger productivity gap.

The wide disparity in estimates of the Japanese productivity advantage underscores the difficulty of making precise calculations, particularly in this context, where the industry structure is different and hard data are relatively scarce. Indeed, precise order of magnitude and the confidence which industry panel members place in the estimated cost difference (i.e., $1200-$1500) comes much more from internal studies, using confidential and proprietary data, than from the relatively rough analysis presented in Appendix A. Some of the calculations in Appendix A are consistent with the internal evidence, and all of them point to a sizeable U.S. cost disadvantage that appears to reflect differences in both prices and productivity.[*]

ASPECTS OF PRODUCT QUALITY

When VW first made significant penetration into the U.S. market, its strategy established a formula for success that has been refined

TABLE 6.1 Estimates of U.S.-Japanese Relative Costs: Alternative Perspectives and Approaches

Category	Industry/Macro	Industry/Micro	Company	Plant
Approach	Estimates productivity and costs using input/output model of whole industrial economy; thus, compares U.S. economy and Japanese; uses published data on wages and input/output matrices.	Examines relative productivity and costs in the auto industry as a whole, including suppliers; uses published data on productivity growth, cost structure, wage rates, and labor hours.	Estimates productivity and costs at the assembler level (Ford, GM, etc.); corrects for difference in product mix and vertical integration; uses data in annual reports.	Productivity and cost estimates based on plant-by-plant comparisons; uses public information on Japanese firms and consultant reports for U.S. firms.
Evidence Productivity differential	Economy wide, U.S. firms have higher productivity ranging from 8 to 20 percent; no estimates given for productivity in auto sector.	On industry basis, Japanese are 20-25 percent more productive.	At the OEM level, Japanese are 41-50 percent more productive.	Depending on specific plant and process, Japanese are 35-200 percent more productive; in plants examined, average is 90 percent Japanese advantage.
Landed-cost difference	Japanese advantage of $1000-$1200.	Japanese advantage of $1400.	Japanese advantage of $1600.	Japanese advantage of $1500.

SOURCE: See Appendix A of this volume.

and extended by the Japanese. A critical element in that strategy was the production of a vehicle that the market perceived to be of high quality. Beginning with the VW Beetle in the late 1950s, the word used most often to describe the character of imported products has been "workmanship," which connotes attention to detail and care in production and quality appearance. The view is now widespread that quality defined in these terms has been a significant factor in the recent success of the Japanese. It seems clear from recent statements of industry executives that an improvement in quality will be an important aspect of any improvement in the U.S. competitive position. In this chapter we identify dimensions of quality that appear to be significant and present evidence about the relative U.S. position.

Definitions

Any attempt to define and evaluate the quality of a complex product such as the automobile must deal with the supposed distinction between perception and reality. It has been argued that the Japanese and European advantage in quality is not "real" in an objective sense but is only a perceived advantage. The implication is that through advertising and other forms of public persuasion the importers have created an image of quality that colors consumer perception. This argument misses the point. In the marketplace, perception is reality. The competitively important dimensions of the product are not those established by experts, nor are the key differences between manufacturers those determined by an objective evaluation. Competitive advantage accrues to those whose products are perceived by the buyers to be of higher quality.

Seen in these terms, quality is simply whatever the market defines it to be. A manufacturer can go to great lengths to offer a car with clearly superior rustproofing, but if corrosion protection is not an element of the market's definition of quality, little competitive advantage will be obtained. There appear to be three dimensions of quality that the market regards as significant. We propose to define each dimension and present some evidence on the relative U.S. position.

Assembly Quality

This category harks back to the notion of "workmanship" identified earlier. It has been described as the "fits and finishes" dimension and includes such things as body finish, squeaks and rattles, the alignment of doors and hoods, and paint quality. Within the industry, assembly quality is usually defined in terms of "building to spec," that is, making the vehicle as specified in the design. This definition focuses attention on the work performed on the assembly line and on the extent to which components and materials meet specifications. But it is clear that the design and thus the specifications themselves are also a significant factor in determining assembly quality. Not only does the design affect the ability of the manufacturing organization to achieve high-quality assembly, but the standards established in the specifications may not be as exacting as those of competitors.

Available information suggests that U.S. producers' assembly quality falls short of the implicit market standard set by the imports. Table 6.2 presents ratings of the condition of selected vehicles at delivery and the number of defects after one month of service. Imports have a clear advantage in both measures, although the gap is largest in defects after one month. These data suggest that consumer perceptions are consistent with actual experience with purchased vehicles.

Reliability

The automobile is a collection of complex mechanical and electrical systems that are subjected to enormous stress--wide variance in temperatures, short bursts of heavy use followed by long periods of inaction, and so forth. To be even minimally competitive, cars must achieve a very high degree of reliability-- that is, the ability to function as designed on demand. In these terms, reliability applies both to individual components and to entire vehicle systems. Failure to function as designed makes the vehicle less useful (at times unuseable), and repairing a malfunction is often a time-consuming hassle. Reliability is thus a critical dimension of overall quality.

To measure reliability we have used the repair incidence data published by Consumer's Union. These data were not drawn from a random sample of all owners but rather from the subscribers to Consumer Reports. They may not be representative of experience generally. The basic data cover repair frequency of mechanical systems and components and the body (structure, finish). Ratings are given in five categories: average, below average, far below average, above average, and far above average. Beginning with a score of 0 for far below average, we have assigned values of 5, 10, 15, and 20 to the other categories. A total score for each vehicle was obtained by summing the scores on individual body and mechanical systems. The results are reported in Table 6.3 for selected makes.

It is apparent that imported products have achieved repair records that exceed those of the domestic manufacturers, in some cases by substantial margins. While the imports have an advantage in both body and mechanical systems, the superiority of the foreign products is most pronounced in the body category. This is consistent with earlier evidence on assembly quality. In mechanical systems, reliability of some of the domestic and imported products is actually quite close. Given the nature of the data and size of the differences, strong conclusions about an overall import advantage in mechanical system reliability does not seem warranted.

TABLE 6.2   Rating of Assembly Quality: U.S. Versus the Imports, 1979

Vehicle Category	Condition of Car at Delivery (scale of 1-10; 10 is excellent)		Condition of Car after One Month of Service (number of defects per vehicle shipped)
Aggregates	*Domestic*	*Imports*	
Subcompact	6.4	7.9	
Compact	6.2	7.7	
Midsize	6.6	8.1	
Standard	6.8	—	
Specific Models			
Omni	7.4		4.10
Corolla	7.8		0.71[b]
Chevette	7.2		3.00
Pinto	6.5		3.70
Rabbit (U.S.)[a]	7.8		2.13
Fiesta	7.9		N/A
Civic	8.0		1.23[c]
Horizon	7.5		N/A
Colt	7.8		N/A

[a] European Rabbit averaged 1.42 defects per vehicle shipped.

[b] Toyota average.

[c] Honda average.

SOURCE: Aggregates: Rogers National Research, *Buyer Profiles*, 1979; Models: Industry Sources.

TABLE 6.3   Ratings of Body and Mechanical
Repair Frequency, 1979
(10 = average; 20 = best; 0 = worst)

Make	Body	Mechanical
Domestic		
Buick	10	10
Chevrolet	4	8
Dodge	8	8
Ford	9	7
Lincoln	10	10
Oldsmobile	11	9
Volkswagen	14	11
Imports		
Datsun	14	11
Honda	16	12
Mazda	18	13
Toyota	17	12
Volkswagen	N/A	N/A
Volvo	16	11

NOTE: The data cover repair frequency of mechanical
systems, components, and body (structure and finish).
Ratings are given in five categories: average, below
average, far below average, above average, and far above
average. Beginning with a score of 0 for those that are far
below average, we have assigned values of 5, 10, 15, and
20 to the other categories. A total score for each vehicle
was obtained by summing the scores of individual body
and mechanical systems.

SOURCE: *Consumer Reports,* April 1979.

Durability

There is little evidence about the long-term durability of Japanese
products and thus little basis for comparison. It does appear that
U.S. products have superior corrosion protection and that basic
components and systems may be more durable.

Customer Loyalty

Perhaps the most significant test of quality production and
customer satisfaction is loyalty, the willingness of buyers to
purchase the same car again. The data presented in Table 6.4

TABLE 6.4 Customer Loyalty: Would Buy Same
Make/Model Again (percentage)

	Domestic	Imported	Total
Subcompact	77.2	91.6	81.2
Compact	74.2	91.4	72.4
Midsize	75.3	94.5	76.9
Standard	81.8	—	—
Luxury	86.6	94.6	87.2
Total	78.7	91.8	—

SOURCE: Rogers National Research, *Buyer Profiles,* 1979.

generally confirm the evidence on assembly quality and reliability. The fraction of owners willing to buy the same make again is much higher in the import group. Since the Japanese dominate the import category, these results underscore their formidable competitive advantage. Not only are their costs significantly lower, but the quality of their products is higher.

EXPLANATION AND PROGNOSIS

It has become almost commonplace to cite the superior quality of Japanese cars as a rationale for their competitive success. With the evidence on productivity and costs, it appears that the Japanese competitive position is buttressed by a significant cost advantage as well as by higher-quality production. On both counts the gap between the United States and Japan is significant. While there is little evidence of a serious attempt to exploit the cost differential through aggressive pricing, it is clear that Japanese manufacturers can absorb very large increases in costs without raising prices and still obtain higher margins than their U.S. competitors.[5] And it is equally clear that a sufficient margin exists for even more costly improvements in performance and quality. Japanese quality levels, however, are already perceived to be a cut above domestic production. With their emphasis on quality and performance the major Japanese firms have acquired a kind of "reputation capital" that enhances an already formidable competitive position.

Popular accounts of the emergence of Japanese producers as first-rate, worldwide competitors tend to emphasize the impact of new automation technology (e.g., robotics), strong support of the central government (i.e., "Japan, Inc."), and influence of Japanese culture (i.e., a dedicated work force). There is no doubt that these factors have played some role. Yet, it is our view that the sources of the Japanese advantage are not to be found in such factors. Rather, they are rooted in a commitment to manufacturing excellence and a strategy that uses manufacturing as a competitive weapon.

It may seem odd to think of manufacturing as anything other than a competitive weapon. After all, "manufacturing" refers to the production and distribution of the product--essential features of competition. Yet the history of the automobile market in the United States suggests that by the late 1950s manufacturing had become a competitively neutral factor. This is not to imply that it was not important; indeed, the Big 3 expended great resources in improving technology and productivity. The point is that none of the major producers sought to achieve a competitive advantage through superior manufacturing performance. Except perhaps for economies of scale, which are affected by manufacturing decisions, the basis of competition was located outside manufacturing--in marketing, styling, and the dealerships.

But the Japanese advantage originates precisely in manufacturing operations. Productivity and quality are determined in the very heart of the operation, in the interaction of people, materials, and equipment. It is in the management of these elements that the Japanese have excelled. And it is the dictates of strategy that have provided the impetus for that excellence.[6]

The strategy of the Japanese producers was first and foremost an entry strategy. The fact that they started from the ground up in the U.S. market influenced their choices in all dimensions of competition. When the Japanese sought to penetrate the U.S. market it would have been foolish to try to compete with the domestic firms on their terms. Just as General Motors (GM) avoided head-on competition with Ford in the 1920s, so the Japanese approach avoided status quo competitive behavior in the 1960s. The domestic market was dominated by the large car, the annual model change, and the "boulevard ride." The new entrants had neither the experience, the production systems, nor the resources to compete on those terms. As with other imports, the Japanese sought out a niche in the small-car segment. Having learned from their early failures (the first attempt at penetration failed on the strength of a low-performance, low-quality product) and the success of VW, both Toyota and Nissan concentrated on establishing a dealer network and on producing a high-quality, solid-performance small car. It was essential that the level of quality and performance be noticeably superior. Otherwise the new product lines were destined to be lost in the competitive shuffle. Moreover, reputation for quality and performance was essential for success over the longer term when entry into higher-margin niches (sports cars, high-performance sedans) was envisioned.

Explaining the Performance Gap

If competitive strategy provides the broad driving force for excellence in manufacturing, what explains observed U.S.-Japanese differences in performance? What aspects of the production process should be singled out for particular notice? To cast some light on these issues, we have identified several characteristics of the production process that may be important in explaining differences in productivity.

The productivity of an operating system--in this case the number of employee hours required per vehicle produced--is determined by the state of technology (both product and process); by the amount and quality of inputs; and by the way in which the resources are combined, organized, and managed. At its most basic level the productivity of an existing operation and technology can be improved either by improving the quality of resource (e.g., hiring more highly skilled workers, using better materials, and so forth) or by more effectively utilizing the existing set. The latter may involve things like changes in supervision, changes in the procedures used to control materials, or a host of other management and organizational factors. Productivity can also be enhanced by introducing advanced technology--new equipment, new products, or new processes and technologies.

These basic determinants--technology, resources, and management systems--can be used to compare and contrast production operations. Our analysis of the U.S.-Japanese productivity gap in auto production is organized around seven factors that have been grouped into three categories: process systems (process yield, quality systems), technology (process automation, product design), and workforce management (absenteeism, job structure, work pace). Any attempt of this sort runs the risk of arbitrary categorization. While useful in clarifying determinants, it should be recognized that many of these are closely related.

Table 6.5 provides definitions of the factors affecting the productivity differential, along with a brief statement of comparative practice in the United States and Japan. The selection of the factors, their definition, and the comparisons are based on discussions with a panel of industry experts. We also asked the panel to rank the factors in order of their importance; some members of the panel provided a percentage allocation. The rankings are presented in Table 6.6.

Perhaps the most striking finding in the panel's assessment is the relative unimportance of the factors connected with technology. Neither automation nor product design is accorded a large measure of explanatory power. Despite the publicity devoted to robotics and advanced assembly plants, such as Nissan's Zama

TABLE 6.5   Factors Affecting Productivity: A Comparison of Technology, Management, and Organization in U.S.-Japanese Auto Production

Factor	Definition	Comparative Practice in Japan Relative to the United States
Process Systems		
1. Process yield	Good parts per hour from a line, press, work group, or process line over an extended period of time; key determinants are machine cycle times, system uptime and reliability; affected by material control methods, maintenance practices, and operating patterns.	Production/materials control minimizes inventory, reduces scrap, exposes problems; line stops highlight problems, help eliminate defects; operators perform routine maintenance; two shifts instead of three leaves time for better maintenance.
2. Quality systems	The series of controls and inspection plans to assure that products are built to specifications.	Japanese use fewer inspectors; some authority and responsibility vested in production worker and supervisor relationship with supplier and very high standards lead to less incoming inspection.
Technology		
3. Process automation	The introduction and adaptation of advanced, state-of-the-art manufacturing equipment.	Overall, state of technology is comparable; Japanese use more robots; their stamping facilities appear to be somewhat more automated than average U.S. ones.
4. Product design	Differences in the way the car is designed for a given market segment; aspects affecting productivity: tolerances, number of parts, fastening methods, etc.	Japanese have more experience in small-car production and have emphasized design for manufacturability (i.e., productivity and quality); newer U.S. models (Escort, GM's J-car) are first models with design/manufacturing specifications comparable to Japanese.
Workforce Management		
5. Absenteeism	All employee time away from workplace, including excused, unexcused, medical, personal, contractual, and other.	Levels of contractual time off are comparable; unexcused absences are much higher in U.S. firms.
6. Job structure	The tasks and responsibilities included in job definitions.	Japanese practice is to create jobs with more breadth (more tasks/skill per job) and depth (more involvement in planning and control of operation); labor classifications are broader; regular production workers perform more skilled tasks; management layers are reduced.
7. Work pace	Speed at which operators perform tasks.	Evidence is not conclusive; some lines run faster, some appear to run more slowly.

TABLE 6.6 Factors Explaining the U.S.-Japanese Productivity Gap: Rankings and Relative Weights from Expert Panel

| | Panel Members | | | | | | | | | | |
| | A | | B | C | | D | E | | Average | |
Factor[a]	Rank	Weight[b] (percentage)	Rank	Rank	Weight[b] (percentage)	Rank	Rank	Weight[b] (percentage)	Rank	Weight[b] (percentage)
1. Process yield	1	30	1	1	30	1	1	40	1	30-40
2. Absenteeism	3	20	3	1	30	2	2	25	2.2	20-30
3. Job structure	2	25	2	5	5	5	4	10	3.6	10-25
4. Process automation	6	6	4	3	15	4	3	15	4.0	6-15
5. Quality systems	7	5.5	5	4	10	6	4	10	5.2	5.5-10
6. Product design	4	7	7	4	10	3	7	0	5.0	0-10
7. Work pace	5	6.5	6	7	0	7	7	0	6.4	0-6.5

[a] For definitions, see Table 6.5 in this volume.
[b] Fraction of the differential explained by the factor.

facility, U.S. firms appear to have maintained comparable levels of advanced process technique and equipment.

The panel's assessment is buttressed by evidence presented in Appendix A that suggests that the Japanese producers may use less capital per vehicle than their U.S. counterparts. While it is true that capital-labor ratios are higher in Japan, the large labor productivity gap cannot be explained by simple capital-labor substitution.

The comparison thus makes clear that an explanation of the productivity gap must be found in the quality of resources and management systems. The panel was unanimous in giving top billing to a factor we have labeled "process yield" but that is really an amalgam of several management practices and systems related to production planning and control. The "yield" category captures Japanese superiority in operating their processes at a high level of good output over extended periods of time. Although engineering (i.e., machine cycles, plant layouts) is of some importance, the key to Japan's lead in this category appears to be the interaction of the material control system, maintenance practices, and employee involvement.

Figure 6.1 graphically portrays the determinants of annual output of good parts (from a representative production process) and indicates some of the management practices and systems that lead to superior performance in Japan. The key to the material control system is the concept of "just in time" production.[7] Often called "Kanban" (after the production cards or tickets used to trigger production), the system is designed so that materials, parts, and components used at a given step in production are produced or delivered just before they are needed. Thus, stages in the process (including suppliers) are tightly coupled, with very little work in-process inventory. Suppliers must therefore make frequent deliveries of parts, and lot sizes must be small to accommodate product variety. It is the Japanese view that reduction of decoupling inventory exposes "the real problems"--waste of time and materials, imbalance in operations, defective parts, equipment operating improperly, and so forth. (Table 6.7 provides comparative data on inventory levels. These data show that dramatically less inventory is used by Japanese firms in the production of automobiles. This is true whether one looks at the process as a whole or at specific plants.) With small buffer stocks the production system will simply not work if there are frequent or lengthy breakdowns. Thus, the just-in-time approach exposes opportunities for reducing waste and solving problems, while at the same time creating pressure for maximizing uptime and minimizing defects. Maintenance programs, preventive and scheduled, are therefore pursued vigorously. Plants operate with only two shifts, and equipment is maintained during nonproduction

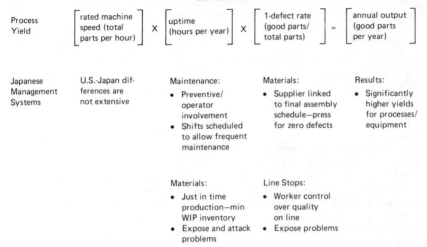

FIGURE 6.1 Japanese management systems and the determinants of process yield.

time. The result is a much lower rate of machine failure and breakdown.

Pressure for defect elimination is reflected in relationships with suppliers and in-work practices on the line. "Just in time" production does not allow for extensive inspection of incoming parts. Suppliers must, therefore, achieve highly demanding quality levels, consistently and reliably. The major Japanese manufacturers work closely with outside vendors to make sure that responsibility for quality is felt and acted upon at the source of product. This same approach--quality control at the source--is used in production on the line, where workers have the authority to stop the operation if they spot defects or other production problems. Worker-initiated line stops are central to the concept of Jidoka: making problems visable to everyone's eye and stopping the line if trouble occurs; all thoughts, methods, and tools to avoid stops are Jidoka.

The basic thrust of the Kanban system and the concept of Jidoka are to eliminate waste, expose problems, and conserve resources. This is not simply a different technique of controlling production, but a very different way of managing the production process. It is clear that these systems interact with other factors in our list of productivity determinants. Separating their effects from the effects of quality systems and job structure, for example, is somewhat arbitrary. The Kanban-Jidoka system uses

TABLE 6.7   Inventory Comparisons—United States and Japan

Level/Process	Japan	United States
1. *Plant and Process Inventories*		
Assembly plant component inventories		
(equivalent units of production)		
heaters	1 hour	5 days
radiators	2 hours	5 days
brake drums	1.5 hours	3 days
bumpers	1 hour	
Front-wheel-drive transfer case in process parts		
storage by operation (number of parts)		
mill	7	240
drill	11	200
ream and chamfer	13	196
drill	24	205
mill, washer, test	10	40
assemble	6	96
finish	7	87
Total	79	1064
2. *Company Inventories*		
Work in process inventories per vehicle		
1979	$80.2	$536.5
1980	$74.2	$584.3
Work in process turns[a]		
1979	40.0	12.1
1980	46.1	13.4

[a] Defined as cost of goods sold divided by work in process inventories.

SOURCES: 1.—Industry sources (data provided by panel members); 2.—Annual reports for representative producers.

fewer inspectors, and its success requires broader and deeper jobs. Seen in this light, the panel's high ranking of "process yield" and the relatively lower importance attached to job structure and quality systems may reflect the difficulty of separating the three factors and a tendency to ascribe to "process yield" some of the impact of the other two.

Indeed, it appears that job structure plays an important role in explaining observed productivity differentials. We have already noted two features of the Japanese system (maintenance practices and Jidoka) in which jobs are designed to involve workers in a variety of tasks. The effects of structure, and the differences in management style and practices that go with it (fewer layers of management, more managing from the bottom up), extend to other aspects of production. Quality circles or "small group involve-ment activities" deal with such questions as layout, process methods, and automation. Such involvement appears to be an important factor in obtaining relatively high levels of commit-ment and motivation.

The nature of worker-management relations in Japan is further suggested by much lower levels of unexcused absence than that found in the United States. The panel accorded absenteeism second billing in its rankings, primarily due to the need to carry redundant workers in order to cover for unexpected absence. Appendix B provides an analysis of this effect. In general, absen-teeism influences costs, not only through redundant labor but also through fringe costs of the absent group as well as indirect effects such as scrap, reduced learning, and so forth. It appears that absenteeism may actually account for as much as 10-12 percent of the cost gap.

Given the impact of absenteeism and the effects of job structure and the workforce influence in "process yield," it is clear that workforce management must be a significant factor in explaining the Japanese cost advantage. Likewise, an attempt to explain quality differences would certainly accord a major influ-ence to the work force and its management. It seems evident, therefore, that in concert with different systems of production management and control, the work force plays a central role in the Japanese competitive advantage.

Much has been made recently of the enormous capital expen-diture programs of the U.S. manufacturers. Indeed, the fact that we have identified the Japanese advantage as a "software" rather than a "hardware" problem in no way implies that new technology could not be effective in raising relative U.S. productivity. Yet it is unlikely that a substantially improved competitive position for U.S. production will be secured only through new bricks and mortar and new machines; comparable processes and machines are avail-able around the world. At least as far as advances in productivity and quality are concerned, new "software" (new approaches to the management of people, materials, and processes) seems essential. The next chapter examines industry experience and prospects in one of these dimensions--workforce management.

NOTES

1. The classical economic theory of international trade posits a world in which trade based on differences in national factor endowments takes place in a competitive world economy. This type of analysis focuses on the long-run equilibrium properties of trade and is essentially static in nature. Recent developments

in the theory of international trade have taken a more dynamic perspective and have introduced innovation, product differentiation, economies of scale, and oligopolistic competition as important elements in determining the pattern of trade. [See the papers by Krugman (1980) and Lancaster (1980)]. The more recent work has focused on explaining intraindustry trade and has incorporated direct foreign investment. Once the assumptions of perfect competition and homogeneous products are abandoned, the notion of comparative advantage becomes more complex. Instead of simple, static comparisons of relative costs of production, which are determined by national differences in factor endowments, the characteristics that determine comparative advantage in the more complex models include differentiation of products, innovative capability, and the nature of domestic competition, in addition to production costs. These characteristics may change over time and may be endogenous. The point is not that relative costs of production are irrelevant, but rather that they are but one element in the determination of a broader notion of comparative advantage and that a full-scale analysis of that broader notion is likely to be complicated. Taken in isolation, relative costs in the foreign country could give misleading indications about trade patterns. Nonetheless, the absence of a comparative cost analysis makes the study less conclusive about the trade pattern than it would be if such analysis had been carried out. Furthermore, a study of comparative costs may put the policy issues in a difference light and would therefore be a useful area for further research. For a recent review of these issues, see Whitman (1981).

2. See Toder (1978).

3. See Katz (1980).

4. Given the differences in estimates of the productivity gap, it is difficult to divide the overall cost difference into a wage portion and a productivity portion. In Appendix A, for example, the industry-level analysis attributes about 20 percent of the cost gap to productivity, while the OEM-level analysis estimates productivity's share to be 38 percent. (These are derived using standard cost-accounting techniques for assigning variances.) In any case, it is clear that both wage and productivity differences are important.

5. There is little evidence that aggressive pricing has been practiced by the Japanese in the U.S. market in Europe; however, the Japanese have priced their products somewhat below comparable domestic vehicles.

6. While the emphasis here is on entry strategies, it should be noted that competition in the Japanese domestic market also had a strong influence on the development of manufacturing capability.

7. For a review of the "just in time" production system, see Monden (1981).

7
Jobs and People:
The Impact of Workforce Management
on Competition

Competition in the U.S. auto industry has undergone fundamental changes in the last 10 years, primarily because of increased penetration of the market by foreign manufacturers. The recent acceleration of this trend has heightened awareness of the changing parameters of competition and raised questions about the viability of domestic production. We have presented evidence that suggests that the Japanese manufacturers have developed a highly productive manufacturing process, a process that produces cars of standard-setting quality. While several elements of the Japanese success are familiar and indeed represent refinements and extensions of concepts and practices developed in the United States, there is a growing recognition that certain, apparently critical aspects of their approach to production are either not incorporated in U.S. practice or have been accorded much less emphasis. Prominent in the latter category is the whole range of policies and procedures connected with workforce management.

While the notion that people are important has been a standard U.S. business cliche for some time, the emergence of the Japanese in world markets in the 1970s has demonstrated both the truth of the proposition and the extent to which U.S. business practice has departed from it. In the auto industry, it is becoming increasingly clear that the work force--the people who design, monitor, maintain, operate, assemble, inspect, supervise, coordinate, and plan--plays a significant competitive role. This role extends not only to the traditional concerns of productivity and cost but also to the quality and performance of the vehicle. It has always been obvious that people were a necessary part of the business, but the Japanese have driven home the point that the work force can be a key weapon in the competitive arsenal. And it is not only in the innate talents of individuals that competitive advantage lies but also in the ability to coordinate and direct those talents, to create a work force capable of outstanding performance.

Some have argued that the competitive advantage of the Japanese (at least as far as workforce management is concerned) is deeply rooted in Japanese culture.[1] In this view, U.S.-Japanese competition is not so much a matter of specific companies' strategies and capabilities as it is a contest of national values, mores, and social goals. While it is true that such factors will influence competition, it seems far too pessimistic to conclude that individual companies have little control over their competitive fortunes. Indeed, the success of several U.S.-based plants of Japanese companies suggests that whatever advantage the Japanese have attained in their work force is less a matter of culture and more a matter of the way the operations are managed.

These considerations suggest that change in the management of people will be a key aspect of any attempt to restore the competitive position of U.S. auto production. Such change must focus on the work force at all levels. While much public discussion of work innovation has been concerned with workers on the assembly line, salaried and managerial personnel and indirect production workers now constitute a large fraction of the industry's work force. Bringing change to the management of salaried and managerial personnel involves a somewhat different set of issues than change among production workers represented by the United Auto Workers (UAW). In this chapter of the report, we focus on the employment relationship at the plant level and in particular on union-management relationships and their impact on workforce capabilities. Many of the issues raised, however, apply (with some modification) to the salaried work force as well. We first trace the impact of technology, unionization, and management practice on the evolution of the employment relationships and then examine a number of innovations that have appeared in the last several years.

TECHNOLOGY AND THE NATURE OF EMPLOYMENT

The employment relationship that prevails in the U.S. auto industry today is the result of a long, evolutionary process that mirrored the development of process technology. We have argued that the course of technology change, at least until 1974, can be seen as an evolutionary process in which a flexible mode of production was transformed into a far more efficient but highly structured form. The evolution from early fluid stages of development to the more rigid specific stage has had a profound impact on the nature and management of work.

In the very early days of the industry the highly skilled, all-purpose machinist played a dominant role in production.[2] Tasks were generally of long duration, requiring a relatively high level of skill. A good deal of art was involved in the metal-work operations, and consequently there was much more reliance on individual responsibility and pride in craftsmanship than on standards or supervision to achieve desired levels and quality of output. Even where supervision was involved, the supervisor tended to be a senior journeyman and the content of the supervision was more likely to be technical and artistic, rather than simply checking up. Personnel policies as we think of them today were rudimentary: the wage structure was chaotic, criteria for job assignment and transfers were ill defined, and hiring was haphazard.

With the emergence of relatively high volume production and consequent standardization and mechanization, production work was transformed. Tasks were significantly reduced in duration and skill content. Machinery operator or tender became the dominant work classification, and dexterity, quickness, and judgement became the dominant skills. The increasing division of labor permitted by mechanization fundamentally altered the relationship between the worker and the product. With tasks highly specialized and fragmented, the connection between any particular individual's efforts and the quality or performance of a recognizable end product became tenuous. Whereas the older, all-around machinist could point with pride to a particular engine block of his work, the engine assembly-line worker who mounted head gaskets had no such ability. In effect, the skills embodied in the old machinist had been transferred to the equipment, and it was the complex equipment rather than the worker that became the subject of wonder and awe.

As the frequency of product change diminished (particularly in the Model T era), as tasks become routinized and the process more mechanized, management practice and the methods of organizational control also changed. The standardization of tasks made possible precise measurement and the development of work standards. Supervision increased in importance as a means of ensuring regular production, adherence to standards, and steady throughput; asset utilization became an increasingly important criterion. To ensure the most effective use of the vast resources embodied in equipment, the auto firms developed an increasingly hierarchical, bureaucratic organization. In managing the large numbers of people at the plant level, for example, the wage structure was rationalized to reflect the skill content of jobs, and policies for hiring and transferring people were formalized.[3] Thus, long before the advent of unionization, an individual worker's employment relationship with the company was structured by rules, standards, and procedures. Most production jobs involved few tasks, and the hierarchical organization structure placed responsibility for control, coordination, and planning of production in a staff group far removed (organizationally) from the actual process.

It is tempting to see the organization and management of production--the locus of responsibility for decisions, the particular methods of control, and so forth--as flowing inevitably from the technology. But this seems to be unjustified. Experience with production in various parts of the world has demonstrated that a given set of machines, processes, and techniques can be managed in different ways. Cultural factors, historical experience, and even management strategy can affect the choice of organizational form. Certainly the technology is important and will create similarities in the nature of tasks and even the design of jobs. But to affect is not to determine. This is particularly true in the "depth" of job definition--that is, how involved workers are in goal setting, planning, coordination, process improvement, troubleshooting, and so forth. It is our view that the emergence of a hierarchical, bureaucratic organization and narrow, shallow jobs in production should be seen as a management choice, conditioned by technology but influenced by a variety of other factors.

While a full analysis of the origins of the management structure and process is beyond the scope of this study, one thing is clear: the organization firms chose to manage the evolving process technology, and the technology itself put a considerable distance between the average worker and the competitive fortunes of the enterprise.[4] The close relationship between the craftsman and the product, a relationship mediated by pride of craftsmanship and involvement in some aspects of design and planning, was gradually replaced by a relationship whose principle nexus was cash.

THE UNION-MANAGEMENT RELATIONSHIP

Changes in technology have continued to influence the character of work and the employment relationship in automobile production, most notably through massive automation in the late 1950s. But the basic direction of change, the move away from artisan skills to operating, monitoring, and maintaining complex pieces of equipment, was established in the Model T era. Since the 1930s, however, collective bargaining has added a new and significant dimension to the process. Unionization added a new institution, and the process of organizing the industry crystallized the views, principles, concepts, and categories of thought that guided the decisions of management and union leaders for years thereafter.

The story of the organization of the auto industry in the 1930s is well known, and its specifics will not be repeated at length here.[5] It was a bitter struggle, characterized by acts of violence and rampant illegal behavior on both sides. The bitterness reflected in part the nature of the times; in part the legacy of arbitrary, heavy-handed supervision and mistreatment on the shop floor; and in part a fear that radical change in power and control would follow union recognition. These perceptions were not completely fanciful. In the 1920s work in the plants had become onerous, especially at Ford, and methods of control increasingly abrasive. Likewise, the demands of the Congress of Industrial Organizations (CIO) and the UAW were radical--a 30-hour week, joint union-management control of line speeds--and the sitdown strike at General Motors (GM) underscored the ability of an organized work force to exert control over operations.

In this context, it is no surprise that Alfred Sloan, Chairman of GM, would resist organization and no surprise that he would refer to the union leaders as "labor dictators." Writing to employees in January 1937 in response to UAW demands, Sloan posed the fundamental issues as he saw them:

> Will a labor organization run the plants of General Motors Corporation or will the management continue to do so? On this issue depends the question as to whether you have to have a union card to hold a job, or whether your job will depend in the future, as it has in the past, upon your own individual merit.
>
> In other words, will you pay tribute to a private group of labor dictators for the privilege of working, or will you have the right to work as you may desire? Wages, working conditions, honest collective bargaining have little, if anything, to do with the underlying situation. They are simply a smoke screen to cover the real objective.[6]

While GM recognized the UAW in 1937, Ford held out until 1941. The reaction of Ford's managers to unionizing efforts was far more vigorous than Sloan's. Members of the Ford Service Department beat up union organizers in the famous "battle of the overpass" in 1937, and suspected sympathizers were fired, harassed, and intimidated. These actions reflected and reinforced the psychological distance between production workers and the firm. In 1941, when Ford finally capitulated, only 5 percent of the workers at the giant Rouge plant voted with the company; 70 percent voted for the UAW, while 25 percent voted for the American Federation of Labor (AFL).

The original demands of the UAW, laid out in 1937, focused on recognition of the union, compensation, seniority rights, establishment of procedures for resolving disputes, and mutual determination of line speeds.[7] Except for the issues of line speeds, all these issues were dealt with in contracts signed by GM in 1937 and by Chrysler and Ford in later years. Line speeds and other production-related matters, such as work standards, the design of jobs, and so forth, were regarded by the companies as management prerogatives, and attempts by the UAW to give the union and workers a larger voice in these matters were strongly resisted by the companies. Over time the union succeeded in introducing work rules and in obtaining influence in the production process through the grievance procedure and administration of the contract. But as far as the individual worker and his daily tasks were concerned, input into decisions about the work and influence over the nature of the job were indirect and limited by management policy.

The organizing effort of the 1930s and the collective bargaining contracts that grew out of it made explicit an adversarial relationship that had begun to form in the interaction of technology and the organization chosen to manage the technology in the early years of the industry. Furthermore, the bitterness of the struggle and the ideology that infused it obviously influenced the character of the new institutions (negotiations, grievance procedures, shop steward/foreman relations) that emerged; the relationship was not only made explicit but also was solidified. As Sloan's statement noted, the basic issue, the principal concern of the companies was not the particulars of wages and working conditions but the fundamental prerogatives of management. These became an end in themselves and the foremen and supervisors their day-to-day guardian.

The employment relationship that developed under the collective bargaining agreements of the 1940s and 1950s was structured by a complicated set of rules and procedures. Compared with the 1920s and 1930s, workers had a greater voice in determining conditions of employment and greater protection from arbitrary action. While the work itself had undergone changes because of increased mechanization and automation, the pressure for production, for meeting budgeted volume and cost objectives, was no less intense. Most public attention focused on the issues of compensation as they surfaced in the periodic negotiations. But the labor-management relationship on the shop floor was dominated by the issues of production--work standards, work pace, attendance, staffing requirements, and so forth.

Although the employment contract became complex, the basic principle seemed to be "a fair day's work for a fair day's pay." Fairness is, of course, relative, and, within the framework of the contract and the grievance process, management used and protected its prerogatives to secure the work it needed to meet its objectives. Standards and supervision remained dominant, and technology aided the control of production through machine pacing. The union sought to protect the contractual rights of its members in the grievance process and negotiated increasing levels of compensation.

Although the adversarial nature of collective bargaining and of the employment relationships persisted and was manifested in occasional strikes and disputes, it seems to have provided a workable approach to workforce management throughout the 20-year period from 1945 to 1965. At the plant level the primary competitive objective was production, an objective well suited to methods of control based on machine pacing, supervision, and work standards. Moreover, workers were relatively well payed, had some control over employment conditions, and had a process at hand for resolving disputes. It was also important that the larger firms in the industry had similar contracts and a similar employment relationship. Differences in style and substance in labor relations existed among the Big 3, but compared with styling, dealership networks, and economies of scale, workforce management was a relatively neutral factor in competition.

The employment contract proved workable in a competitive environment that emphasized production and acceptable quality and with workers familiar with the depression and World War II. As conditions and people changed in the late 1960s, it became increasingly unprofitable. In retrospect it is clear that, while the traditional employment relationship gave the worker a limited voice in setting conditions at work and while it paid well, it did not secure loyalty of commitment. Nor was it intended to. Neither the competitive situation nor managerial organizational strategy placed a premium on loyalty or commitment. Production objectives required people to operate the equipment, but as long as minimum standards were met (and supervisors ensured they were) loyalty or commitment was not essential.

Changes in the work force in the late 1960s and Japanese and European competition in the 1970s placed new demands on the employment relationships. The problems of younger workers, the issues of alienation, boredom, and aversion to production work have been well documented.[8] In effect, these problems indicate the extent to which the employment relationships relied on forms of compensation not as highly valued by a new generation. These developments provided a source of internal pressure for change. At GM, for example, concern about the performance of the workers led to a range of studies and projects in organizational development in the late 1960s and early 1970s. But even at GM, innovations in workforce management diffused very slowly in the organization.

In contrast to the lack of substantive change in the employment relationship before the 1970s (despite all the popular attention afforded the "blue collar blues"), the current competitive crisis seems to have convinced management leaders of the need for change [union leaders (national level) have supported changes in workforce management for some time]. The competitive thrust of the Japanese has altered perceptions, while it has altered the standards of comparison in the marketplace. A rethinking of the competitive value of the traditional relationship is under way.

Given the new competitive environment, the traditional adversarial relationship can be unprofitable both because of the practices and attitudes it engenders and because of those it does not. We have already noted the lack of any incentive for loyalty or commitment in the bargain. Clearly, high quality requires some measure of care in production. It has apparently been assumed that assembly quality could be insulated from worker influence by dividing tasks, by automating, by closely supervising the work, and by inspecting extensively. While acceptable quality can perhaps be achieved in this manner, outstanding production requires something more.

It is important to realize that even if workers were committed and loyal and sought high quality the organization of work creates significant obstacles. With an emphasis on production, with supervisors under pressure to keep the line moving, a worker has little incentive (to put it mildly) to be more careful than the minimum requires or to check previous operations. Indeed, the pressure of the line often has precisely the opposite impact.

The traditional employment contract relies on supervision and standards to ensure an acceptable volume and quality of production. But it is also true that it relies on the willingness of the worker to be subject to the supervision and the standards. Furthermore, the grievance procedure creates an opportunity for the union to exercise a degree of influence on the production process. If there is not some kind of agreement on short-term objectives, the grievance procedure can be used to play all sorts of games. A typical pattern is for the shop steward to build up a backlog of grievances (some justified, some not) to be used as bargaining chips for reduced workloads that plant management buys off in order to avoid disruption. At the same time, supervisors, under pressure to get production and improve efficiency, harass people to get results, which creates opportunities for more grievances. This kind of vicious cycle is not harmless politics; it can have serious effects on quality and productivity.

Another problem with the traditional relationship, one that is often overlooked, is the failure of management to tap the information and experience embodied in the work force. Suggestion programs notwithstanding, a basic assumption of the bargain and one incorporated into practice is that management knows how to make the cars, and the worker's job is simply to follow instructions. Experience has shown, however, that workers have valuable information and insight that can be profitably employed. An example from GM's Tarrytown plant--a relocation of trim departments--illustrates the point:

> At first, the changes were introduced in the usual way. Manufacturing and industrial engineers and technical specialists designed the new layout, developed the charts and blueprints, and planned every move. Two of the production supervisors asked a question that was to have a profound effect on events to follow. "Why not ask the workers themselves to get involved in the move? They are experts in their own right. They know as much about trim operations as anyone else." Old timers in the union report wondering about management's motives. Many supervisors in other departments also doubted the wisdom of fully disclosing the plans.
>
> Nevertheless, the supervisors of the two trim departments insisted not only that plans not be hidden from the workers but also that the latter would have a say in the setup of jobs. The supervisors were impressed by the outpouring of ideas: "We found they did know a lot about their own operations. They made hundreds of suggestions and we adopted many of them."[9]

THE POTENTIAL FOR CHANGE

It is now generally recognized among the leaders of both the companies and the UAW that the old employment contract is increasingly unprofitable. The process of change has begun, and, as the Tarrytown experience illustrates, considerable progress has been made. The magnitude of the adjustment is substantial. If this brief discussion of a complex problem leaves any impression, it should be that the employment relationship--the relationship of the company, the union, and the worker--is a capital resource of the firm. It is a long-lived asset. The character of the relationship today is the legacy, in part, of decisions made long ago. To change the management of the work force is to change not only rules and procedures or organizational structure but also the habits of mind, patterns of interaction, and whole categories of thinking.

Like any capital asset, change requires investment. What "investments" in changing the employment relationship and workforce management are under way in the industry? What is the potential for successful change? We have already noted the efforts at GM to introduce organizational development and "Quality of Work Life" programs in production operations. This effort began almost 10 years ago and has spread in one form or another to almost 80 percent of GM's plants. The approach has been "bottom up," with managers of facilities in the critical decision-making role. Thus, while there is pressure for plant managers and local union officials to have some kind of program, specific approaches and initiatives are not imposed from the top.

An effort to create "employee involvement programs" is under way at Ford. Working with the UAW leadership, and using a combination of grass-roots and "top down" methods, Ford has made the need for high quality an organizing theme for its efforts. Programs to create a climate of involvement and commitment are in progress at well over half of Ford's facilities. Significant strides have been made in developing a more cooperative relationship, with measurable improvements in product quality a tangible result.

In league with the UAW, both Ford and GM have begun to redefine the nature of the employment relationship and the structure and organization of work at the plant level. Success will require not only change at the level of the production worker but also in middle management and at the very highest levels in the organization. Indeed, research at GM and elsewhere has shown that even at the plant level, improvements in workforce management and performance require a commitment and leadership from the top plant and union officials and involvement and organizational change among staff and line middle managers.

A similar conclusion about commitment applies to the corporation as a whole. Recent changes at GM are instructive. Two top-level appointments were recently made that may signal new attitudes and approaches to quality and labor relations. For the first time ever, GM has placed a senior-level executive, in this case a vice-president, in charge of quality. Furthermore, GM's new Vice-President of Labor Relations was drawn from the personnel side of the business and has been involved and deeply committed to GM's programs in quality of work life. These developments seem to indicate a recognition of the need for continued organizational innovation.

Prospects for Adaptation

The organizations that produce automobiles in the United States are immense. The attitudes and practices, the patterns of thought and action that underlie employment, are deep-seated. How likely is it that such large organizations will succeed in adapting and changing in such fundamental ways? We offer no definitive prediction, but a number of key factors can be identified. Our research on productivity and organizational change in other contexts suggests that successful instances of adaptation share four characteristics; these are spelled out in Table 7.1.

The available evidence suggests that the impetus for successful change comes from both internal and external pressures. When one kind of pressure is absent, firms are less likely to engage in a fundamental examination of the operation of the enterprise. That kind of rethinking of the basic premises of the organizational strategy appears to be a key aspect of successful change. Unless management undertakes a thorough review and calls the basic operating procedures into question, change is likely to be superficial.

Review of procedures is more likely to be fundamental and far-reaching if it is led by a newcomer with influence and expertise. The newcomer may be a consultant or a line manager who appears to have the advantage of fresh perspective and few internal political connections. The final characteristic spelled out in Table 7.1 is the style of decision making and problem solving. The evidence suggests that successful change occurs in the context of shared power and authority, when those affected by the decision have a voice in its determination.

Table 7.1 presents both an ideal situation in this framework and an assessment of the auto industry's position along these dimensions. It is by now abundantly clear that the industry faces extraordinary internal and external pressures. Whereas declining profitability has been a source of concern over the last 15 years, the question is now one of survival (at least for U.S. operations). Looking at other factors, the industry has begun what appears to be a thorough examination of the workforce issue. Further, there is some evidence in both GM and Ford that the approach to change in workforce management involves a sharing of power between bosses and subordinates; the close involvement of the UAW is also supportive of this view. Finally, there is little evidence of the role that "newcomers" might play, but there is some indication of a search for and a receptivity to new ideas, new approaches, and people to back them.

There should be no misunderstanding. The factors sketched out in Table 7.1 are not a recipe, but rather a set of conditions that appear to be important in instituting organizational change. Nor should the magnitude of the task facing the industry be underestimated. In the case of productivity, product quality, and the role of the work force, we are talking about something close to a cultural revolution, about fundamental changes in the way the business is managed and the ways that people at all levels participate in the enterprise. Recent developments suggest a good deal of adaptability in the collective bargaining relationship and thus some reason for optimism.

TABLE 7.1 The Auto Situation and Characteristics of Successful Organizational Change

Characteristic	Ideal Type	Situation in the Auto Industry
1. Pressure on management	Management faces pressure for improvement of performance from both internal (organization) and external sources—example: ineffective working relations between different levels of management (internal); decline in market position—low profitability (external).	Worker/union concern and discontent over workforce management provide internal pressure; vigorous competition from the Japanese imposes strong external pressure.
2. Catalytic intervention	A newcomer (an outsider to organization) enters in a position of influence; the outsider brings expertise relevant to current problems.	Some new managers introduced; new awareness/receptivity to alternative approaches to workforce management.
3. Thorough reexamination of operations	Newcomer encourages a fundamental reexamination of past practices; assumption that management knows the answer is discarded.	Recognition of the need for fundamental change in the work force exists; continuation of efforts begun in early 1970s; basic premises of workforce management are undergoing reexamination.
4. Shared authority/power	Problems defined and/or solved through group discussions and implementation; power is shared between bosses and subordinates; organization members given voice in decisions that affect them.	Approach to change largely "bottom up"; close involvement of UAW and local leadership.

SOURCE: Adapted from Kim Clark, *Unions and Productivity in the Cement Industry* (Unpublished Ph.D. thesis); see also L. Greiner, "Patterns of Organizational Change," *Harvard Business Review*, 1967, pp. 119-130.

NOTES

1. A particularly clear statement of this view was articulated in the 1980 Fiat Annual Report.

2. The importance of the all-purpose machinist has been amply demonstrated by Rae (1959); see especially Chapter 3.

3. This process has been documented by Chandler (1964).

4. This is obvious in the history of such groups as the Ford Security Service; see Chandler (1964), pp. 215-218.

5. See, for example, Fine (1969).

6. Chandler (1964).

7. The history of the contract has been documented. See, for example, Monthly Labor Review (March 1937), pp. 666-670.

8. Some of these have been overblown. For some interesting insights into work on the assembly line, see Guest (1973).

9. Guest (1979), p. 78.

Technology and Competition in the U.S. Automobile Market

The crisis in the U.S. auto industry reflects in part the vigorous competitive challenge of the Japanese, in part a rapid shift in market preferences. We have argued that a recapture of competitive cost and quality performance will require fundamental changes in the way the manufacturing process in the U.S. auto industry is managed. At the same time the industry is faced with the problem of developing new products to meet changing market demands. "Small" and "fuel efficient" are the words most often used to describe desired characteristics, and, however superficial those descriptions may be, it is becoming increasingly clear that competitive vehicle design requires product technology different from that found in the standard American sedan of the post-World War II period. Throughout that era, competition in any given segment of the U.S. automobile market occurred largely on the basis of economies of scale, styling, and sales and service networks. As befits a maturing industry, innovation became increasingly incremental in nature and, in marketing terms, virtually invisible. Has that situation now changed?

Clearly, some major changes in product technology have occurred in the last few years, most notably the introduction of front-wheel drive, the trans-axle, and the increased use of the diesel engine. Yet it is not clear whether such developments reflect the beginning of new technological ferment or the end of a technological transition. Indeed, some suggest that the small, front-wheel-drive car, with its transverse mounted, four-cylinder engine, already constitutes a new dominant design. If so, future changes in automobile technology are likely to be incremental to the new design, and competition will occur much as before on the basis of styling, scale economies, and dealer networks. This reading of events places the industry farther along its path toward maturity.

If, however, innovation in technology is again becoming of vital competitive significance, not only are we likely to see continuing functional innovations but the overall pace of innovation is likely to increase as well. These developments, in turn, may have far-reaching implications for the structure of the industry, for individual strategic decisions, and for the pattern of international trade. The question of technology's role in competi tion is thus central to interpreting current events and is an important aspect of the larger question of the industry's future.

In marketing terms the notion of "competitive significance" has both a quantity and a price dimension. A given technology (e.g., front-wheel drive) becomes significant in a competitive sense either if consumers are willing to pay a premium for cars embodying the technology or if the market share of such models increases. Accordingly, our analysis in this chapter will focus on both sales and price effects of a few major technological characteristics in the context of a fairly simple model of consumer demand (at the compact and subcompact end of the market). Our hypothesis is that the oil shock of 1979 altered demand patterns, thus increasing both the visibility of technology and its competitive significance. If true, we should observe very different market valuations of those characteristics before and after 1979.

A FRAMEWORK FOR ANALYSIS

The framework we use to test the competitive significance of technology is designed to identify the market's valuation (sales and price) of a given characteristic.[1] There are few difficulties with sales effects. In fact the only major issue is the problem of capacity constraints. In a given year the sales of a particular model may be more a reflection of the capacity of the firm to produce it than of underlying consumer demand. But since our purpose is to estimate the average effect on sales of a given characteristic, and since each characteristic tends to be found on more than one model, it follows that only if all models with a given characteristic are subject to capacity constraints will sales reflect those constraints and not market demand.

The sales model we use is presented in Appendix C. Our approach simply is to identify the impact of a particular technological feature (e.g., front-wheel drive) by statistically holding constant the effects of other characteristics.

Estimating the market value attached to specific characteristics is a somewhat trickier process, for it relies on a set of assumptions that must be spelled out clearly. The problem here lies in the fact that there is no market for individual characteristics as such. A given model contains by definition a bundle of characteristics or attributes, and its price in the market is the price of the bundle as a whole.

To identify the "price" of a particular attribute, we assume that observed prices are a reflection of an implicit market in characteristics. Consumers place a value on specific attributes that have been aggregated or bundled in different ways to form different models. To infer the value of a specific characteristic, we take a range of model prices, some of which have the characteristic and some of which do not. By observing how the overall price changes with variation in characteristics, we can identify the value consumers place on specific attributes. The details of this approach are presented in Appendix C.

A difficulty arises, however, in interpreting the prices of these attributes. What we are after is the market's or the consumer's valuation of each characteristic, but the market price of a given model is determined by the interaction of supply as well as demand forces. On the supply side the key determinants are the costs of production, the pricing policy (e.g., relationship of one model's price to another), and the strategy of the firm. The demand side is determined by the preferences of consumers, i.e., by their assessments of the value of a given characteristic. Though only demand considerations are relevant for our purposes, the price of a model reflects both market valuation, which interests us, and the manufacturer's costs and pricing policy, which do not.

One solution to the problem of isolating the demand effects is to specify a structural model of consumer demand and the costs of production. Given appropriate exogenous variables one can use advanced statistical techniques to disentangle supply and demand effects.[2] But a solution much less demanding of the data may be available. The lack of identifiability of the demand function may not affect inferences about changes in demand. Costs of roduction may be more stable than consumer valuation, particularly over short periods of time when the sales mix of characteristics shifts rapidly. If this is true, then any differences in coefficients estimated for two closely paired years will reveal the influence of changes in consumer demand. Furthermore, the availability of both list and transaction prices may provide another means of correcting for supply side effects.

Existing evidence on automobile pricing suggests that list prices are determined by product-line policy and standard costs.[3] While the markup over standard cost may be influenced by strategic considerations and estimates of consumer valuation, variations in list prices are likely to reflect variations in cost. Inclusion of the list price in an equation explaining transaction prices may therefore provide a control for the supply side. It is of course possible that such a procedure will "overcontrol" and thereby obscure demand effects. But estimation with and without the list price in a comparative context should provide a basis for inferences about shifts in demand.

Empirical Specification

The framework developed above requires data on sales, list and transaction prices, and model characteristics that capture the effect of technology and innovation. List prices and sales data are readily available for most models sold in the United States, but transaction prices for new cars are not. As a proxy for them we have chosen to use the price of the model one year later as determined in the used-car market. Use of the year-old price introduces a further complication in the analysis, since discounts based on used-car prices will reflect the effects of deterioration as well as the market's valuation of technological character-istics. However, as long as deterioration is not a function of those characteristics, it is fair to assume that the effects of general deterioration will be reflected in the overall average for the market and will not obscure relevant model differences.

The characteristics we have chosen to include in the analysis are determined by our hypothesis about changes in the role of technology in the market for automobiles. To begin, we distin-guish between performance characteristics and technological attributes, although both are closely related. Thus, miles per gallon measures fuel-efficiency performance, but the size of the engine is a characteristic of the model, and both kinds of vari-ables are included in the analysis.[5]

In the compact and subcompact markets on which we focus, the key performance characteristics are fuel efficiency, driving range, repair frequency, and package efficiency (efficient use of space). The relevant definitions and measures are presented in Table 8.1. The key technological characteristics include engine type (gas or diesel), drive train (front or rear wheel), and age. This last variable is meant to test the notion that newness itself is valued independent of specific characteristics.

We are aware that this scheme leaves out, by necessity, a number of alternatives that may be important in the market. Since the omitted factors may be reflected in the analysis if they are correlated with variables that are included, care must be taken in interpreting the result. The drive train, for example, may pick up some of the effects of differences in handling and maneuverability.

To complete the framework we have added a set of variables that control for the country of origin and the market segment. Variables indicating whether a car is produced in Japan (we distinguish captive imports from others), Europe, or the United States are intended to pick up any otherwise unmeasured differ-ences in quality of attributes correlated with the country of origin. Finally, we have allowed the average discount to be different in the subcompact and compact model categories.

TABLE 8.1 Basic Variables: Definitions

Variables	Symbols and Definitions
Sales	S_i: number of vehicles of ith model sold in specific year.
List price	P: list price.
Transaction price	P*: one-year-old used-car price.
Fuel efficiency	MPG: EPA miles per gallon rating for city driving.
Driving range	RNG: MPG (reported fuel-tank capacity).
Repair frequency (two variables entered)	REP1: has value of 1 if *Consumer Reports'* survey of repair frequency placed model in "substantially below average" category; zero otherwise. REP2: has value of 1 if *Consumer Reports'* survey of repair frequency placed model in "substantially above average" category; zero otherwise.
Package efficiency (two measures)	IVTV: internal volume (defined below) ÷ total volume (length × width × height). VOLWT: internal volume ÷ vehicle curb weight.
Engine type	DIESEL: has value of 1 if model has diesel engine; zero otherwise.
Drive train	FWD: has value of 1 if model has front-wheel drive; zero otherwise.
Age of model	AGE: years since last major model redesign.

NOTE: Using the diagrams and specifications in the figure below (taken from *Consumer Reports' Annual Auto Issue,* April 1980), the internal volume is obtained by adding the internal volume of the front and rear compartments of the car. The internal volume of the front compartment is the product of the front shoulder room and the cross-section area of the front compartment. The internal volume of the rear compartment is similarly obtained. The formulas are as follows:

Front compartment volume V1 = J $[(0.5) (G - 18)^2 + (24) (E) + 0.18 (E)^2]$
Rear compartment volume V2 = K $[(H - 18) (G - 18) (0.71) + (F) (H)]$
Internal Volume = V1 + V2

The following assumptions were made in the calculations:
- distance between tester's hip and knee = 18 inches
- tester's leg was inclined at 45°
- seats were inclined at 10° to the vertical

The dimensions E and F were defined as the height above seats rather than the clearances measured by *Consumer Reports'* tester. *Consumer Reports'* E and F were modified by adding a constant of 34 inches to obtain the heights above seat levels. Data for the E adjustments were obtained from *Automotive News Almanac.*

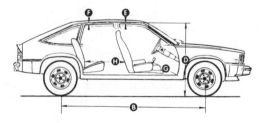

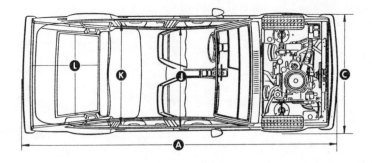

Table 8.2 presents average values of the basic variables included in our analysis of sales and discounts. A comparison of 1977 and 1979 data reveals that most characteristics are quite similar in the two years. Because of the introduction of new models and substantial redesign efforts, the average age fell from almost four years to a little over two and one-half.

Among the other characteristics, however, we find little difference. The percentage of models with front-wheel drive increased slightly, as did driving range. But the changes are small, and in general the average characteristics are similar in the two years. The model-age data suggest, however, that substantial changes may have taken place both in the way in which the technology was packaged and in the quality of the technology.

It is important to realize that the subcompact market in 1977 included models (in comparable numbers and at comparable levels of performance) offering the characteristics we hypothesize were relevant in the 1979-1980 market. The data suggest that a comparison of market results in these two years might be a useful test of the role of technology in competition. Had we found gross dissimilarities in attributes, the validity of the test would have been suspect.

COMPARISONS OF DISCOUNTS DURING 1977 AND 1979

To compare the impact of given attributes on the transaction price (statistical results for discounts are presented in Appendix C, Tables C.3 and C.4), we have developed what we call the basic effect of each characteristic. We first calculate how much a "typical" difference in the characteristic would have raised or lowered the price.[5] For VOLWT (package efficiency), for example, a typical or average difference between a given model and the average for all models was 8.7 cubic inches per pound in 1979. We estimate that such a difference would raise the

TABLE 8.2 Means and Standard Deviations of Basic Variables, 1977 and 1979

	1977		1979	
Variable	Mean	Typical Difference From the Mean	Mean	Typical Difference From the Mean
Sales per (S) model (000s)	92.6	93.4	102.8	95.1
List price (P)	4484	1209	5657	963
Transaction price (P*)	3674	901	4591	832
Fuel efficiency (MPG)	21.8	5.3	21.0	4.8
Driving range (RNG)	318.0	57.7	326.5	70.8
Package efficiency				
VOLWT (cubic inches per pound)	49.0	7.0	49.3	8.7
IVTV (percent)	22.5	1.7	22.3	2.8
Engine type (percent diesel)	2.3	14.9	3.5	18.5
Drive train (FWD) (percent)	25.0	43.3	27.1	44.4
Age of model (AGE)	3.7	3.5	2.3	2.1
Repair frequency (percent)				
REP 1	15.9	36.6	3.5	18.5
REP 2	22.7	41.9	15.3	36.0
Subcompact (percent)	65.9	47.4	50.6	50.0

SOURCE: *Automotive News Almanac,* 1978, 1979; *Consumer Reports* (annual auto issue 1978, 1980).

transaction price by about $30. This is the basic effect. In the case of characteristics that are either present or absent (e.g., diesel engines) the basic effect is the effect of having the characteristic. To find out whether the basic effect is a large or small one, we compare it to the typical difference in transaction prices, which in 1979 was $832. Thus, the basic effect of VOLWT ($30) was equal to 3.6 percent of the typical difference in the transaction price in 1979. We call this the relative effect.

It may be useful to restate our basic hypothesis in terms of the variables in the analysis. In the price analysis the issue of competitive significance is essentially a question of the size and magnitude of the basic and relative effects. If a given technology characteristic, e.g., front-wheel drive, is a positive factor in competition, we would expect models with that characteristic to be higher priced. Thus, the sign of the basic effect should be positive; of course, an important factor will have a larger relative effect.

Table 8.3 contains the basic effect and the relative effect of technological characteristics for 1977 and 1979. The results are striking. A line-by-line comparison of the effects of the various characteristics reveals a sharply different pattern of market valuation in these two years. While MPG was a positive impact in 1977, its effect in 1979 is swamped by the technology variables, and the estimated fuel-economy effect turns negative. The negative effect of MPG is puzzling but may reflect collinearity with the technology variables. Collinearity may also affect estimates of parameters on the technology variables. This seems to be especially true for VOLWT (package efficiency), where a large standard error precludes any strong conclusions about the sign of the price effect in either 1977 or 1979. For FWD (front-wheel drive) and AGE, however, collinearity does not seem to be a serious problem. If, for example, we drop the variables for diesel engine, age, and package efficiency (VOLWT) from the analysis, leaving only front-wheel drive as a measure of technology, we obtain basic and relative effects of front-wheel drive very similar to those reported in Table 8.3. And even if we reduced the 1979 FWD coefficient by two standard deviations, the resulting estimate (111) is still positive. In general, we find clear differences between the results for 1977 and those for 1979.

TABLE 8.3   Basic and Relative Effects on Price Discounts of Performance and Technology Characteristics, 1977 and 1979[a]

Characteristics	1977		1979	
	Basic Effect (current dollars)	Relative Effect (percentage)	Basic Effect (current dollars)	Relative Effect (percentage)
Performance Characteristics				
Fuel efficiency (MPG)	192	21.3	−67	−8.1
Driving range (RNG)	5	0.6	65	7.8
Package efficiency (VOLWT)	−57	−6.3	30	3.6
Technology Characteristics				
Engine type (DIESEL)	−853	−94.6	302	36.3
Drive train (FWD)	−316	−35.0	417	50.1
Model age (AGE)	−40	−4.4	−85	10.2
Market Segment				
Subcompact (SUB)	−174	−19.6	313	37.6

[a] Effects for 1977 and 1979 are based on the coefficients given in equation 6 of Table C.3 in this volume.

The evidence suggests, for example, that consumers were willing to pay a premium for a diesel engine over and above premiums for greater range and greater fuel efficiency. In contrast, consumers in 1977 heavily discounted cars with front-wheel drive and diesel engines. Moreover, even the performance characteristics valued in 1979--range, package efficiency--had small or negative effects in 1977. While the market placed a high value on fuel efficiency in 1977, the performance and technology characteristics associated with market success in 1979 had little market appeal in 1977. If features that were considered innovative in 1979 had been introduced in 1977, they would have been greeted with above-average discounts. Introduction in such a context could only be interpreted as an attempt to force the market, for it seems clear that innovation, at least as defined in 1979 terms, was not valued in 1977.

Perhaps the clearest indication of the value of innovation in 1977 and 1979 is the effect of model age. In 1979 the basic effect of a typical difference in age (2.3 years) was to lower the price by $85. Thus, newer cars received a premium even after controlling for other attributes. Since performance and technology are already accounted for, the result implies that newness per se carried a premium in 1979.

The effect of age in 1977 has the same sign but is much smaller and, as Appendix C shows, is not statistically significant. The results on the age variable in 1979 stand in contrast to the patterns of competition that have prevailed in the auto industry during most of the postwar era. Older models tend to be debugged, refined, and developed to meet a relatively stable set of consumer demands. A major model change may introduce new features and above-average performance, but if innovation is not itself valued the effects of bugs and introduction problems are likely to outweigh any value attached to the new features. This situation reversed in the late 1970s.

The notion that market valuation of performance and technology was different after the oil shock of 1979 can be examined more rigorously using standard statistical tests. The null hypothesis in this connection is that the effects in the two years are identical. Appendix C provides technical details, but the statistical tests confirm the apparent differences in Table 8.3. We are able to reject with a high degree of confidence the hypothesis that the effects are equal. In short, the evidence suggests that market valuation changed sharply after the oil shock of 1979, with technology playing a much more critical role.

COMPARISON OF SALES PER MODEL DURING 1977 AND 1979

The effects of performance and of technology characteristics on model sales in 1977 and 1979 are examined in Table 8.4. In contrast to the results for price discounts, we find remarkable stability in the pattern of effects for 1977 and 1979. For most variables the direction of the effects are identical, and in several cases the orders of magnitude are roughly comparable. Even where there are differences the evidence in Appendix C suggests that little should be made of these results. The quality of the statistical evidence is poor, and the estimated effects are not statistically different from one another.

Thus, while the more negative effects of diesel and front-wheel drive cast doubt on the hypothesis, and the evidence on age supports it, these changes are in fact more apparent than real. The data available provide no evidence that technology (as defined

TABLE 8.4 Basic and Relative Effects on Sales of Performance and Technology Characteristics, 1977 and 1979[a]

| | 1977 | | 1979 | |
| | Basic Effect (thousands | Relative Effect | Basic Effect (thousands | Relative Effect |
Characteristics	of vehicles)	(percentage)	of vehicles)	(percentage)
Performance Characteristics				
Fuel efficiency (MPG)	35.0	37.5	10.1	10.6
Driving range (RNG)	−46.2	−49.5	−6.6	−6.9
Package efficiency (VOLWT)	32.9	35.2	36.5	38.4
Technology Characteristics				
Engine type (DIESEL)	−0.6	−0.6	−97.8	−102.8
Drive train (FWD)	−23.4	−25.1	−53.1	−55.8
Age (AGE)	4.9	5.2	−1.8	1.9
Repair Record				
Much worse than average (REP 1)	7.9	8.5	−61.3	−64.5
Much better than average (REP 2)	82.2	88.0	65.2	68.6
Market Segment				
Subcompact (SUB)	−84.8	94.0	3.0	3.2

[a] Effects are based on the coefficients given in equation 3 of Table C.4 in this volume.

here) had a positive effect on sales in either year or even that the sales relationships shifted dramatically. It may be that market shares are more stable than prices and that most of the effects of market shifts in 1979 were felt on the price side of the market. The results of the price analysis provide evidence that technology was an important aspect of competition in the 1979-1980 market. Compared with the situation in 1977, we find that technology characteristics and innovation were highly valued. At least as far as market premiums are concerned, changes in the market created an incentive for innovation consistent with the hypothesis advanced at the beginning of this chapter. In terms of their effect on prices, technology became more visible and competitively significant.

NOTES

1. There is a large literature on the estimation of hedonic price equations. A review of the literature and an application can be found in Toder (1978).

2. In effect what is required is a structural model of the implicit market for characteristics. With appropriate exogenous variables and exclusion restrictions, the model can be identified and estimated using one of several simultaneous equation methods.

3. It is obvious that this is a strong assumption, since the firm's pricing policy may attempt to estimate consumer valuation. Thus, the mode is only an approximation.

4. Although identification of estimated discounts effects depends on the assumption that list prices are based on standard costs and standard markups, the change in the estimated effects between two periods is likely to be more affected by demand considerations since relative costs of characteristics tend to be more stable.

5. This allows us to identify the effect of the technology-holding performance constant.

6. In statistical terms, "typical difference" is the standard deviation of the variable in question.

7. Evaluation of the results should be tempered by the fact that the standard errors (in Appendix C of this volume) are sizeable in some cases. The lack of precision precludes strong conclusions for some variables. However, the overall pattern of effects is what is important, and it appears that overall the two patterns are different.

8. Discount equations with average repair records also were estimated, but the results were insignificant and were not reported in the test; see Appendix C of this volume.

9

The Character of Automotive
Innovation in the 1970s and Beyond

The rising cost of oil is the driving force behind the current ferment in technology in the auto industry. From 1973 to 1980 the real price of gasoline rose over 80 percent. In 1979 alone, the nominal price more than doubled. The market is now demanding levels of fuel economy significantly in excess of the mandated corporate average fuel economy (CAFE) standards. The clear competitive advantage accruing to products with advanced efficiency performance has created an incentive for the development of improved hardware. If the real price of oil continues to rise and we experience significant supply interruptions, the future course of product innovation may become more radical.

Questions about the character of innovation and technology are thus closely linked to questions about the future of oil prices. Given a national commitment to reduce reliance for oil on the Middle East, it is of course possible that regulation could force innovation even without any change in oil prices. But market forces are likely to be far more conducive to radical innovation than regulation. It is thus essential to examine the likelihood of continuing increases in the real price of oil and gasoline.

Experience in the last two years reveals the large degree of uncertainty surrounding movements in the price of oil. Not only is the world price set through a complex political and economic process--a process subject to swift and radical developments--but we have had little experience on which to base judgements and estimates.[1] Though organized in 1960, OPEC did not exercise significant control over the price of oil until the embargo of 1973. The shift in power in 1973 following the war in the Middle East was sudden and pervasive; the oil companies, long used to a negotiated price, were faced with ultimatums and drastically higher prices. The sudden jump in world oil prices and the price of gasoline marked the end of a long period of stability. Moreover, the impact of the events of 1973-1974 was partially cushioned by

133

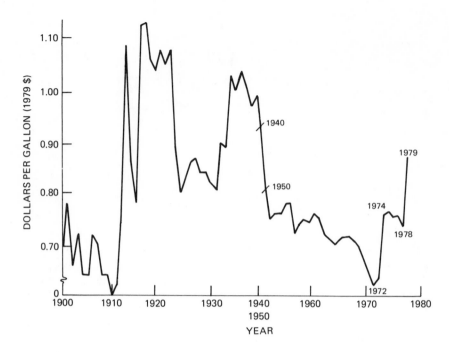

FIGURE 9.1 Real price of gasoline from 1900 to 1979 (in 1979 dollars).

government regulation in the United States. Both developments are illustrated in Figures 9.1 and 9.2, which present historical data on gasoline prices in the United States and comparative data for other countries around the world.

Because the full impact was not immediately felt, because the weight of history was on the side of stability, the 25 percent jump in the real price of gasoline was widely viewed as temporary. In January 1975, Business Week published an article on OPEC that carried the subtitle "The 12-Nation Club Is Powerful Today--But May Have to Lower Prices Later." The article contained several quotes that reinforced that theme. The statement of Richard Gonzalez, a Houston oil economist, is representative:

> You can almost bet that OPEC has overshot the mark on the price of oil for the long term. They'll find out the real equilibrium price of oil in the 1980s will be less than it is today. It is possible that OPEC will see the development of alternatives coming and will ease off on the price of oil to keep it from getting out of hand.[2]

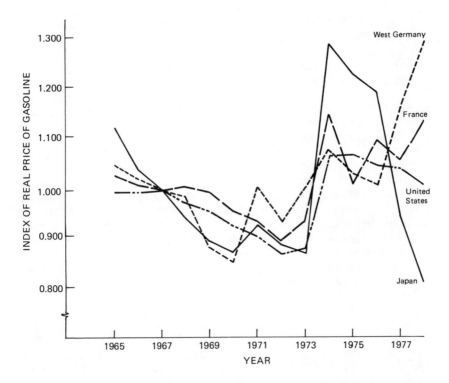

FIGURE 9.2 Real price of gasoline in selected countries, 1965-1978 (1967 = 1.000). (Data for 1965-1975 from International Road Federation; for 1975-1978 from U.S. Department of Energy, National Petroleum News Factbook, 1979.)

As the figures show, developments in 1975 and 1976 seem to reinforce this view, as the real price of gasoline fell in most of the major auto-producing countries. Price controls in the United States helped to extend the period of slowly falling real prices into 1978. Predictions of steady real prices of gasoline were common. Thus, it is no surprise that the congressional Office of Technology Assessment in its projections of auto technology published in 1978 chose to use $1.20 per gallon as the price for gasoline for the year 2000. Although some observers, notably the Central Intelligence Agency, took a far more pessimistic posture, these were dismissed by persuasive arguments. Writing in the fall of 1978, Congressman David Stockman argued against a national energy policy, declaring that "the global economic conditions necessary for another major unilateral price action by OPEC are not likely to re-emerge for more than a decade--if ever."[3]

Subsequent developments in 1979 laid to rest such notions and revealed the fragility of the world oil market. Political forces, both in broad international terms and in terms of Middle East politics, have assumed a major role in the long-term development of the supply and price of oil. The revolution in Iran and the Iran-Iraq war have underscored two critical aspects of the integration of politics and economics in oil. The first aspect is the sensitivity of the market to relatively modest shifts in supply. The shutdown in Iran in 1979 removed about 10 percent of world supply, yet this shortfall precipitated a scramble that drove spot market prices to $40 per barrel. Secondly, the Iranian revolution demonstrated the tension between tradition and modernization inherent in oil-dominated Middle East development strategies. Second thoughts about the wisdom of high production rates appear to be wide-spread; the argument is that oil in the ground may be a better long-term social and economic investment than western real estate or massive building programs.

On top of the obvious political instability in the Middle East, these considerations suggest that the real price of gasoline will continue to increase over the long term and that periodic explosions in prices and disruptions in supply are to be expected. It is also true that this long-term scenario is likely to be accompanied by periods of stable or declining real gasoline prices. OPEC pricing seems to involve a ratchet effect, in which a given burst in the nominal price is followed by a period of relative stability until the next major increase. Unless more of the world's oil comes to be traded on the spot market, this pattern is likely to continue.*

The prospect of a series of price bursts followed by periods of stability may create problems for the U.S. auto industry. On the one hand, they must plan and develop technology consistent with long-term demands, but they may face periods in which fuel efficiency becomes relatively less important and other factors (recreation, comfort) dominate sales. If, however, the events of 1979 have altered perceptions of the long-term situation, long- and short-term demands may become consistent. At least for now there appears to be ample evidence that the old days of relatively cheap gas are gone forever and that market pressures for fuel-efficient vehicles will persist.

DIVERSITY AND RADICAL CHANGE IN TECHNOLOGY

From the introduction of the Model T in 1908 to the oil embargo of 1974, innovation in the auto industry was conditioned by and reinforced a convergence in products and processes. Earlier we documented the increasingly incremental character of technological advances and the implications of such innovations for strategy and competition. Spurred by the transformation of the oil market and shifts in demand, the pattern seems to be changing. Our analysis of the market's evaluation of technology and innovation in the late 1970s suggests an increased role for technology in competition. Casual impressions suggest an increasingly diverse array of automotive hardware in engines, drive trains, and other systems. There are indications that innovation is less oriented toward refinements of existing design concepts and focused on the development of new approaches. Yet the picture is not clearly defined.

Considerable resources are being and have been allocated to the development of a technology that can only be described as incremental. While this state of affairs is to be expected if the industry is in a period of transition, better evidence than casual impressions is required to establish the character of technological changes and to provide a basis for judging future developments.

Evidence must be developed through an examination and interpretation of specific innovations; our approach is to compare recent developments with the historical pattern and with innovations known to be "in the wings" but not yet introduced. We are concerned with the general pace of innovation as well as its general character. Two aspects are especially critical. The first is the diversity of technology growing out of the innovative process; the issue is essentially whether, for any given system, a new dominant design is apparent. The second aspect is the extent to which innovation departs from design concepts currently in use, whether innovation is epochal or incremental.[5]

The issues of diversity and dominance in design require a relatively straightforward analysis of technology features embodied in current production and identification of apparent trends. Classifying a particular development as incremental or epochal (or something in between), however, is a more ambiguous problem. To sharpen the distinction and to establish criteria for classification, it may be useful to sketch out the framework of analysis more fully than we have done.

Whether an innovation is radical or incremental is essentially a question of perspective. What is radical from one standpoint may be only incremental from another. A change in engine technology, for example, may be greeted as a great step forward by the driver of the car but might have only minor effects on the equipment and people in the production process. In the current analysis we have chosen to adopt the perspective of the production processes. Thus, the key issue is how a given change in product technology affects existing capital equipment, labor skills, materials and components, management expertise, and organizational capabilities within the production unit. Note that these categories encompass the full range of activities involved in the development, production, and marketing of the product.

From the perspective of the production unit, a truly epochal innovation is one that destroys the usefulness of existing competence in several of the factors of production (capital, labor components, management, organization). Such an innovation sweeps through the production process, leaving obsolescence in its wake.[6] Perhaps the most striking example of an epochal innovation was the introduction of closed steel bodies in the 1920s and its impact on Ford and its Model T. Confronted with a shift in market preferences and competition from firms producing closed bodies in great variety, Ford was forced to totally revamp the production process, replace 15,000 machine tools, introduce new processes, and lay off and hire thousands of workers. Moreover, management skills and organization appropriate to the production and marketing of a low-price standardized, mass-produced automobile were not viable in the era of the annual model change and increasing variety.

Contrast Ford and the closed steel bodies of the 1920s with the introduction of the thin-wall, gray cast-iron engine in 1959.[7] Improvements in metallurgical consistency of gray cast iron and the development of more accurate mold fabrication allowed Ford to reduce the engine wall thickness from 0.20 inches to 0.15 inches, which increased thermal efficiency in addition to reducing weight. This was an important development for Ford because it allowed the company to compete with the new compact cars equipped with aluminum engines using a familiar technology. Thus, far from making existing capabilities obsolete, the thin-wall engine preserved Ford's investment in cast-iron technology and associated labor, managerial, and organizational skills. The new technology extended and refined an existing concept, and innovation was incremental.

Between the extremes of such epochal developments as closed steel bodies and the truly incremental changes such as the thin-wall engine, there is a wide range. Our analysis will attempt to place specific innovations on the spectrum, but we shall not make fine distinctions; definitive categorization would require fairly detailed information beyond the scope of this study. In addition to assessing particular innovations, we shall attempt to gauge possible interdependencies between innovations. Such interaction has been of significance in the past, as a cluster of innovations emerges that reinforce one another to produce major changes, even though any particular innovation may, in isolation, have been only marginally important. In the case of interdependence, as with an innovation considered singly, the critical issue is the number of factors of production that must be transformed if the innovation is to be actually used.

Diversity and Dominance in Design

The historical course of innovation in engines and bodies has been marked by a succession of what we have called "dominant designs," i.e., particular design approaches that evoke noticeable competitive reaction and that result in significant market penetration. George White has argued that dominant designs can be recognized early in their development.[8] One or more of the following attributes appear to characterize designs that will achieve dominance:

* Technologies that lift fundamental technical constraints limiting prior art while not imposing stringent new constraints.
* Designs that enhance the value of potential innovations in other elements of a product or process.
* Products that assure expansion into new markets.

It should be noted that a dominant design is not typically the product of radical innovation.[9] To the contrary, a design approach becomes dominant, as did the integration of engine plants with transfer lines and the closed steel body, when the weight of many innovations tilts the economic balance in favor of one design approach. Typically, the relevant design approach has already been in existence. It may appear radical in a particular application, because the newly favored concept replaces a much different alternative, but the competing approaches were probably the product of evolutionary improvement.

The importance of evolutionary change is evident in the cars introduced just after World War II. In almost every major system (engine, transmission, body, etc.) the designs could be traced to a series of developments dating back some 15-20 years.[10] The concepts embodied in the large road cruisers of the late 1940s and early 1950s proved to be dominant one in the U.S. market for over 20 years. It is true that some movement away from the basic configuration took place as early as 1958, but the orientation of the market toward ample power and a smooth, luxurious, boulevard ride did not change markedly until the late 1960s and early 1970s.

The first movement away from the all-purpose road cruiser was signalled by penetration of the market by imports in the late 1950s. At that time the Corvair's systems and features were already a significant departure from the industry norm, and additional changes (aluminum engines, unit body construction) were broadly introduced at various points until 1969. None of these developments (except perhaps for unit construction) displaced the dominant postwar designs, but they indicated the beginning of divergence in market-tested configurations.

Throughout the 1950-1969 period there appears to have been a logical, evolutionary shift in the locus of innovation. Technical change was oriented toward meeting new competitive objectives while preserving existing capabilities. The response to the initial import surge of the late 1950s was generally to scale down existing designs. Pollution regulations were first met with add-on components and modifications of existing systems. Similarly, initial gains in fuel economy were realized by changes in components and not in basic design.[11]

As problems with imports and environmental regulations have persisted and have been compounded by higher fuel prices, however, the chain of ramifications has extended further up the design hierarchy to affect the basic configuration of the car. Separate frames and bodies in smaller cars have been replaced with a rationalized design combining the two in unitized construction. Body designs have been changed to reduce the number of parts. The recent developments of transverse-mounted engines and front-wheel drive strike at more basic relationships among components.

A similar pattern is apparent in engine design. The search for greater efficiency has been met with refinements of the basic technology, such as that of Nissan's NAP-Z engine--a move to fewer cylinders, redesign of the combustion chamber, use of exhaust gas recirculation, and so forth. In addition, the turbocharger (an add-on device) has been used by Saab and Buick as a way to maintain performance while reducing engine size. Its most promising operation, however, may be with diesels. The diesel engine offers an alternative concept of increasing popularity; a shift to diesel technology in passenger cars constitutes a refinement of an existing technology. More fundamental changes in propulsion systems are under development.

The thrust for improved efficiency reflects a series of trade-offs among fuel economy, emissions, and safety. An example illustrates the incentives and the constraints facing producers. The easiest and cheapest way to improve fuel economy is to make the car smaller, thereby decreasing weight without adding innovative technology. There are limits to this approach, some imposed by marketing considerations but more important ones by safety concerns. As cars become smaller, the point is reached where there is not enough "crush distance" to limit the g-forces on a car's occupants if there is a crash. So one turns to innovative technology: new structural materials to make the car lighter without making it smaller and new engine and drive-train systems to improve efficiency at a given weight. In a sense, these technologies have a safety objective and may require a premium price.

The development of product technology in the 1970s constitutes a sharp reversal of the pattern of technical change that dominated from 1900 to 1950.[12] In that period, innovation moved toward a standardized design. The process of standardization followed a hierarchy: first came the propulsion choice, then the overall chassis configuration, and then major components were advanced. Finally, once technological change in the components subsided, the overall design of the automobile was optimized. In the aftermath of energy shock and shifts in market preferences, the hierarchy has been reversed. Change came first to components, then to the overall body, and now changes are appearing in the drive trains and engines.

A comparison of the technology currently on the market with that available in 1973 reveals the diversity that recent innovation has spawned. The market now offers both gas and diesel engines; engines with four, five, six, and eight cylinders; engines with computer-optimized control; and engines with turbochargers. One can buy front- or rear-wheel drive; downsized and/or redesigned bodies with new lightweight, high-strength materials; and new kinds of automatic transmissions. Yet in the midst of this diversity there appears to be at least some focus on ongoing developments, and some technologies appear to be achieving significant market acceptance. To illustrate this point, it is useful to consider the technology embodied in the new Ford Escort and the Chrysler K car. Both were designed in the last few years, and both are using their technologies as key selling points; it seems reasonable to expect that they would be representative of the most recent trends in technology.

Although these cars are aimed at different segments of the market, their technological features are similar. Both offer a transverse-mounted, four-cylinder engine with overhead camshaft and aluminum head. The K car's engine design is (in the words of Car and Driver) a "bit archaic" in its cylinder head, stroke length, and combustion chamber; the Escort has a compound valve semi-head configuration and specially designed pistons to improve combustion. The drive trains of both cars are packed into the front of the vehicle (both have front-wheel drive) and both have a four-speed manual transmission as standard equipment, although the Escort has overdrive in fourth gear. Rack and pinion steering and front disc brakes are standard on both.

When these cars are compared with the market leaders of 1979-1980, two conclusions are clear. First, front-wheel drive with a transverse-mounted engine seems to be on its way to achieving market dominance in the small, economy-car segment. The popularity of the transverse-mounted/front-wheel-drive configuration seems to come from its interaction with downsizing and vehicle redesign. While front-wheel drive may offer superior handling characteristics, it also seems to provide greater package efficiency than the rear-wheel-drive format. The second point is that a car with four cylinders seems to be the configuration of choice. Such a car seems to provide the right mix of weight reduction and preservation of performance.

It is important to note in this context, however, that a good deal of the current R&D effort seems to be focused on developing alternative power plants.[13] The existence and likely future availability of a small diesel option with turbocharging is further evidence of the lack of an overall dominant design in the engine. Thus, while four cylinders seem to be a focal point, additional (and more critical) engine characteristics are in flux.

Incremental versus Epochal Innovation

The changing locus of technological development (components, chassis, propulsion system) in the 1970s suggests that innovation is becoming less incremental in its impact on the production unit. Although thoroughly sweeping changes have yet to be introduced, there is ample evidence that the industry is in the midst of a technological transformation that may have profound implications for the basic factors of production. Tables 9.1 and 9.2 illustrate the kinds of innovation experienced in the postwar era as well as future developments that appear to be under active investigation.

Looking first at the evidence in Table 9.1, the preferences of the market are clear in the growing shift from performance to efficiency as a basic objective of innovation. Most of the major product developments from 1945 to 1974 were designed to improve handling, increase power, and generally improve the recreational value of the product. Safety regulations and greater demand for a more efficient vehicle have become increasingly important since the mid-1960s; performance-oriented changes have almost disappeared. The shift in objectives underscores the fact that innovation has been significantly affected by market demand. And if the future developments given in Table 9.2 are at all indicative of the actual course of technology in the next several years, the efficiency-oriented pull of the market will be strongly felt for some time to come.

Whether the objective is performance or efficiency, the impact of any one particular development may be magnified if it influences developments in other areas. Historically, individual innovations have solved specific problems or added new features, but they have seldom been independently decisive in causing one approach to dominate its competitors. Often a shift in market performance coupled with an innovation causes one approach to gain in preference over another, as with the shift toward small cars and bodies of unit construction. Because improvements are cumulative, the chance decreases with time that a single

TABLE 9.1 Characteristics of Major Technological Innovations, 1946-1980

Innovation	Date	Objective[a]	Potential for Interaction[b]	Diffusion[c]	Impact on Production Unit[d]
Engines					
1. High-compression V-8	1949	Performance		72 percent by 1955; over 85 percent by mid-1960s.	Extended existing capabilities; dominant design.
2. Aluminum engine	Late 1950s	Efficiency		Maximum of 10 percent in 1961.	Moderately significant; represents new design concept in material; may obsolete some equipment and skills.
3. Stratified-charge engine (CVCC)	1975	Efficiency/ emissions		Limited at present.	Extends existing technology; old four-, six-, and eight-cylinder engines can be refit with system.
4. Turbocharging	1977-1978	Efficiency/ performance	Affects downsizing options; engine design and control; emissions technology.	Available as an option on several models but increasing.	Add-on device at present; if brought in-house, adds new requirements.
Drive Train					
1. Automatic transmission	1948	Performance		70 percent by 1955.	Moderate; requires new facilities.
2. Disc brakes	Mid-1960s	Performance/ safety		Standard equipment on front brakes by mid-1960s.	Moderately significant; new design concept—obsolete manufacturing equipment and skills; has achieved dominance.
3. Transverse FWD/trans-axle	Mid-1970s	Efficiency	Affects downsizing and design of interior.	FWD available on 25 percent of small cars; growing predicted 50 percent penetration by 1985.	Requires new components and makes some equipment and processes obsolete; substantial impact.

TABLE 9.1 (continued)

Innovation	Date	Objective[a]	Potential for Interaction[b]	Diffusion[c]	Impact on Production Unit[d]
4. Four-speed automatic/TCLU	1979-1980	Efficiency	Cannot be used on FWD cars.	Limited.	Some new tooling and equipment, moderate impact.
Body Systems					
1. Energy-absorbing steering assembly	1966	Safety	Not significant.	Standard equipment on all cars by 1968.	New materials and design in shaft; requires some new equipment, but has only minor impact on existing capabilities.
2. Downsizing	1973	Efficiency	Interacts with engine design; drive-train configuration; materials.		Requires significant capital investment but does not obsolete existing process or skills.
3. Materials substitution	1973	Efficiency	Affects downsizing and interacts with safety technology.		Heaviest impact (at present) on suppliers; could obsolete existing technology (e.g., switch from steel to plastic).
4. Electronics	1978	Efficiency	Highly significant in engine design and control; transmission, emissions (future developments even more significant).		Add-on devices and some in-house production; impact has been moderate, potential impact more significant.

[a] The objective is the primary purpose/need that the innovation was intended to serve/meet; we have distinguished between efficiency in operation and that in performance.
[b] The potential for interaction is what influence(s) the innovation will/does have on others (e.g., makes them more or less feasible/necessary).
[c] Diffusion is the extent to which an innovation in use at a particular time; the pattern of adaption.
[d] For example, is the old process made obsolete or simply extended? It is moderately affected or not at all? Is a whole new facility required or simply some new tooling?

TABLE 9.2 Characteristics of Selected Future Technologies, 1980 and Beyond

Innovation	Date	Objective[a]	Potential for Interaction[b]	Impact on Production Unit[c]
Engines				
1. Direct injected stratified charge (e.g., PROCO)	1983	Efficiency	Strong interaction with electronics.	Moderate; requires some new feature but does not obsolete existing equipment, skills, etc.
2. Advanced diesel (e.g., adiabatic turbocompound)	1985+	Efficiency	Interacts with turbocharging and turbines; requires new materials.	Could be extensive; a new design concept requiring new materials and emphasizing three technologies (diesel, turbocharging, turbines) in combination.
3. Electric vehicles	1984-1985	Efficiency	Interacts with downsizing; materials.	Extensive; new energy source; new propulsion system; if widely adopted, would obsolete existing engine manufacturers.
4. Turbines	1985+	Efficiency	Interacts with vehicle design, materials; also drive train (continuously variable transmission); emissions technology.	Extensive; new propulsion system, multifuel capability.
5. Stirling	1985+	Efficiency	Technology interacts with materials.	Extensive; new components and materials; multifuel capability.
6. Stratified charge rotary engine	1985	Efficiency	Interacts with electronics, fuel-injection systems.	Extensive, little commonality with reciprocity engine.
7. Flywheel hybrid	1985	Efficiency	Significant interaction with CVT and engine design.	Moderate (relatively low energy densities; therefore, conventional materials).

TABLE 9.2 (continued)

Innovation	Date	Objective[a]	Potential for Interaction[b]	Impact on Production Unit[c]
Drive Train				
1. Continuously variable transmission	1982+	Efficiency	Significant interaction with engine design.	Significant, new skills, equipment, technique.
Body/Systems				
1. Materials (e.g., structural plastic)	1980+	Efficiency	Significant interaction with engine design, drive train, body configuration; affects downsizing options.	Could be extensive; shift to plastics/ceramics-based technology could obsolete existing capability.
2. Electronics	1980+	Efficiency/safety and performance	Extensive interaction with all systems.	Through interation could be a gradual influence (could have a major impact on nature of product.
3. High-pressure tires	1983	Efficiency	Interacts with suspension system and brake design.	Negligible.
Alternative Fuels				
1. Methanol	1985+	Efficiency		Could preserve existing systems; can be used with conventional engine configuration.
2. Hybrid fuel cell	1985+	Efficiency	Could be significant.	Potentially radical; based on chemical principles, could obsolete existing systems.

[a] The objective is the primary purpose/need that the innovation was intended to serve/meet; we have distinguished between efficiency in operation and that in performance.

[b] The potential for interaction is what influence(s) the innovation will/does have on others (e.g., makes them more or less feasible/necessary).

[c] For example, is the old process made obsolete or simply extended? It is moderately affected or not at all? Is a whole new facility required or simply some new tooling?

innovation will change a favored approach. Significant change arises as the result of a cluster of interacting innovations.

Recent innovations in materials and electronics appear to offer significant potential for interaction. In addition to being an important development in its own right, electronic sensing and processing of information is an "enabling" technology; without it a variety of developments in engine design (e.g., advanced diesels) would either never have advanced or would have advanced slowly. With it, however, not only might individual technologies advance more rapidly, but innovations in separate systems can be linked. A potentially important example is the linking of a continuously variable transmission with an advanced engine concept (e.g., adiabatic engine) to achieve significant increases in efficiency. Thus, while electronics may not directly affect the production unit in a radical way, its indirect impact may be extensive.

In general, it is clear from Tables 9.1 and 9.2 that radical change is a likely consequence of technology currently "in the wings." Note that this conclusion applies only to the impact of innovation on the production process. It does not refer to the scientific or technical novelty of the innovation. Indeed, the discussion in Chapter 2 showed that a relatively minor technical change, such as downsizing, can have a significant effect on the production unit. In this sense, it appears that innovation in the auto industry has become increasingly less incremental. Downsizing, for example, does not make existing stamping and assembly plants obsolete, but it is a process different from the annual model change. Indeed, in some parts of the production unit (product design, materials), downsizing has not simply extended existing technology; rather, it has required new concepts. Similarly, the trans-axle requires extensive changes in axle and transmission plants and, from that perspective, constitutes a more radical change than would refinement or extension of rear-wheel drive and conventional transmissions.

The evidence suggests that innovation in the 1970s generally has proceeded first where the cost of change (in terms of its impact on the existing process) has been least. This serves to underscore the potential for change in future years. The technologies in Table 9.2 involve not only new design concepts but also in many cases totally new physical or mechanical and chemical principles. And indications are that such developments are not the flight of some engineer's fancy; extensive development work is under way in all areas and is some cases has been speeded up remarkably in the last two years. During 1982, for example, reports of important developments in battery technology have opened up the possibility of electric vehicle commercialization within the next five years.

There are two points to note about such future developments. The first is the obvious point that most of these innovations have the potential for transforming important segments of the production unit. Were the electric car to dominate vehicle production, for example, a large part of the existing engine manufacturing process, including labor skills and management expertise, would be obsolete. Furthermore, the propulsion technology may interact with other systems (transmission, chassis) to produce further changes. Other innovations listed in Tables 9.1 and 9.2 might have equally profound effects.

The second point is that, at least for engine technology, many of the designs are competitive; obviously, it is as yet unclear which will dominate. Moreover, it is not clear that even prototype development will serve to indicate the extent of market acceptance. Since the innovations tend to be destructive of existing capital, and since they themselves require large amounts of capital for development and production, there appears to be significant risk associated with these developments. Indeed, the range of uncertainty about future technology appears to be growing. Dealing with that uncertainty, positioning the organization for adaptation and management of change will be critical to survival if technology-based competition becomes a reality.

Yet the potential payoffs are significant. The implication is that incentives for innovations in engines, materials, and other technologies are strong; that it is unlikely that all producers will choose the same line of development; and that it is equally unlikely that any given producer will pursue development in all areas. Depending on the nature of technical breakthroughs, it is entirely possible that the market will see a diversity of advanced-design power plants and other systems and components as the available options compete for market acceptance; in terms of product technology, a period of intense technological competition may be just ahead.

NOTES

1. This section draws on the work of the Energy Project at the Harvard Business School. See Stobaugh and Yergin (1979a,b).

2. "OPEC: The Economics of the Oil Cartel," Business Week, January 1975, pp. 80-81.

3. Stockman (1978).

4. As of early 1982 real oil prices had fallen for several months.

5. See footnotes 1-4 in Chapter 3 for sources of innovation.

6. This is the kind of innovation to which Shumpeter referred when he coined the phrase "winds of creative destruction."

7. See Abernathy (1978), pp. 211-212.

8. White, G. (1978).

9. The application of this concept to the auto industry and to conceptual development can be found in Abernathy (1978, see Chapters 2-4). This section draws extensively on his analysis.

10. Abernathy (1978) documents this position for both engines and bodies.

11. Basic design in this context refers to such design concepts as front-wheel drive, the principal of energy transformation (gas, electric), and so forth.

12. Abernathy (1978; see Chapter 3, which documents the pattern of change up to 1970).

13. See Heywood and Wilkes (1980).

The Automotive Future:
Three Scenarios and Their Implications

At the beginning of this report we sketched out three interpretations of the current crisis in the U.S. auto industry. Although we have not provided a detailed or exhaustive analysis of each, we have focused on two aspects of the situation--technology and comparative costs--that distinguish the three lines of interpretation from one another. The various assumptions made and the evidence developed are summarized in Table 10.1. The evidence points to several things: a sizeable cost and quality disadvantage for U.S. producers, some new designs that appear to have widespread appeal, and a general ferment in technology that foreshadows what may be radical innovation over the next several years.

Table 10.1 underscores the ambiguity in the current setting. As might be expected in a period of transition, we have identified evidence that accords with some of the assumptions of all three competing explanations. Since the purpose of this report is to highlight possibilities, we draw no strong conclusions. Indeed, it is important to note that the actual course of development may not be monolithic. Maturity may characterize the industry for some time, followed by a period of fundamental change. Furthermore, one pattern of development may not describe all market segments.

In light of the uncertainty clouding the future and in order to clarify the implications of alternative patterns of development, this concluding chapter of the report presents three scenarios of the industry's future, based on the three lines of interpretation. We will sketch out conditions and interactions that are consistent with the course of industry activity implied by a particular interpretation of the current crisis. We will discuss the management implications and the implications for alternative public policies for each scenario. The scenarios depict possible chains of events and the likely impact on those events of broad public policy options. While the scenarios are intended to offer a realistic assessment of the development of the industry under given

TABLE 10.1   A Summary of the Three Interpretations and the Evidence

	Assumption by Category of Interpretation			
	Transient Misfortune	Natural Consequences of Maturity	Fundamental Structural Changes	Available Evidence
Competitive Position				
Cost	U.S. costs on par or slightly above competitors.	U.S. firms have substantial disadvantage based on factor prices.	Makes no strong assumptions; (implicitly similar to maturity level.	Major competitors (Japanese) have a sizeable cost advantage ($1000-$1500) based on wage rates and productivity.
Quality	U.S. products are comparable.	Quality not a critical element in competition.	No strong assumption.	Quality is central to competition, and U.S. firms are at a disadvantage.
Technology and Competition				
Role of innovation	Neutral	Neutral	Significant advantage accrues to innovators.	Market values newness and new technology; premiums paid for hardware.
Diversity of technology	New dominant design achieved.	Designs standardized.	Technology increasingly diverse; no dominant design.	Diversity has increased; some features/systems have achieved significant acceptance (four cylinders; front-wheel drive).
Impact of innovation on productive unit	Incremental after new designs established.	Incremental changes preserve existing competence.	Increasingly radical; destructive of existing competence.	Appears to be less incremental and more radical; future developments could radically change process but are uncertain.

assumptions, they are not based on an extensive analysis of business strategy. And although some general views about public policy are indicated, an in-depth analysis of policy options was not carried out. The strategies of particular firms and detailed policy analysis are important areas for further work but were outside the scope of this report. We draw no conclusions about the desirability of specific government actions; rather, we indicate how alternative actions may affect the industry within a given scenario.[1]

THREE SCENARIOS

The three interpretations of the current crisis--transient economic misfortune, natural consequence of maturity, and fundamental structural change--have different implications for the future. Any attempt to grasp those implications, however, must recognize a point that is often missed: the usual form of forecasting the future, trend projection, is often misleading. The future typically emerges from a collision of events that are often unrelated. For example, the oil crisis of 1979 would have had little permanent effect on the U.S. automotive industry if it were not for three other events: (1) Japanese penetration of the U.S. market and their development of excess capacity, (2) the U.S. government's earlier attempt to shelter consumers by controlling oil prices and mandating fuel-economy standards, and (3) the U.S. auto manufacturers' misjudgment of the future.

If any view of the future is to offer useful insight it must be more than a linear projection of simple trends; it must recognize interactions among shifting technologies, political events, and economic factors. One way to address this complicated reality is through the development of scenarios. A scenario is a depiction of possible chains of events that might occur. Their use provides insight into the various ramifications of changes in associated conditions.

The three scenarios are presented in Figures 10.1, 10.2, and 10.3. No attempt has been made to describe the course of events in great detail. Furthermore, we have assumed no changes in current government policy. Our purpose is to indicate a few main lines of development that are apt to occur under a given set of assumptions. Thus, the scenarios illustrate the consequences of different assumptions, taken jointly.

The basic assumptions of each line of interpretation can be summarized as follows:

1. Transient Misfortune: Small or negligible cost/quality disadvantage.

Central problem is lack
of small-car capacity.
New dominant design has
emerged.

2. Consequences of Maturity: Product and process
technologies are stable.
Large cost disadvantage.
Location of production
based on factor prices.

3. Fundamental Restructuring: Technology becomes
competitively important.
Continued increases in
real price of oil.
Introduction of radical
innovations.

Scenario 1: Transient Misfortune

The implications of the view that the industry is in the midst of a serious, albeit transient, misfortune are spelled out in Figure 10.1. The left column depicts some of the driving forces in the scenario: (1) protectionist sentiment growing out of declines in auto sales; (2) the mismatch of U.S. car capacity and market demands; and (3) the emergency of a broader diversity in technology and product mix. Where before V-8s had powered a largely similar fleet of U.S.-produced cars, suddenly there is an expansion of models at the bottom of the market (e.g., Toyota's Starlet) and "at the sides," with repect to the propulsion options already introduced or in process as well as with respect to other technological features.

From this starting point, Figure 10.1 shows a snowball effect, as the implications of new capacity and a weakened domestic industry unfold. U.S. producers develop new models and small-car capacity, but the extra Japanese capacity leads to intense price competition. Some of the new capacity from Japanese firms comes from U.S.-based assembly plants. The U.S. firms, with their attention focused on the most popular models, fail to meet Japanese competition in emerging niches and segments; the mini-car is an example.

Although the depths of misfortune are transient, they clearly have lasting effects. U.S. firms regain some market share but overall are not financially strong enough to completely recoup their losses, nor are they able to offset Japanese penetration into new segments. The picture is brighter on a national basis. New Japanese assembly plants and new capital invested by the U.S. firms leave the country with a viable domestic industry. Because

154

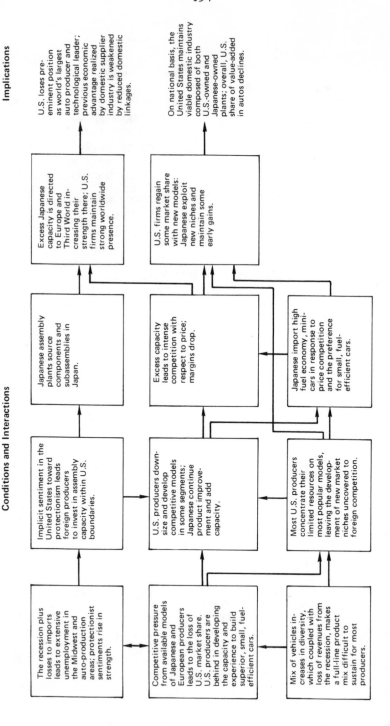

FIGURE 10.1 Transient economic misfortune.

of market-share losses by domestic firms and the tendency for Japanese plants to source in their home country, the U.S. share of value-added declines, and the industry decreases in overall size. The United States loses its preeminent world position, and the domestic supplier industry is weakened. The net effect is a smaller, but viable, domestic production base.

Scenario 2: Consequences of Maturity

The second scenario differs primarily in the assumption of a significant U.S. cost disadvantage and a highly stable product and process technology. In our lexicon a "mature" industry is one in which the technology of product and process is essentially embodied in factors of production that can be readily purchased in established markets. Assuming that the auto industry is mature in this sense has major ramifications. As Figure 10.2 suggests, the current crisis is but a continuation of a long-term restructuring of the location of production.

A known and stable technology and wide cost differences imply that most cars sold in the United States will be produced in such countries as Taiwan, Brazil, or Mexico, where labor costs are much lower than in the United States and even lower than in Japan ($1 per hour in Taiwan versus $9 per hour in Japan and $18 per hour in the United States). This scenario also implies an increasing role for suppliers who can move their plants rapidly. As events unfold, this search for low-cost production leads to a loss of production capability and know-how in the United States and an increase in third world countries. Cars produced in these countries are introduced into U.S. markets.

This scenario does not assume that vehicle production moves completely offshore. Because of engineering expertise, unique market demands, and other factors associated with product differentiation, it is likely that a good fraction (e.g., 35 percent) of the cars sold in the United States will be produced domestically. Added to the decline in production by the original equipment manufacturers (OEMs) is the loss in creative, innovative interaction among the auto and other industries and the loss in manufacturing know-how in high-volume vehicle manufacturing.

Scenario 2 includes three forcing events depicted at the left-hand side of Figure 10.2: (1) the effects of the recession in reducing demand; (2) the continued inroads of Japanese producers based on their landed-cost (and quality) advantage; and (3) a change in consumer preferences for efficient transportation modes. The implications of the first two events are rather direct: standardization, worldwide sourcing, foreign penetration. The third event means a more competitive and more varied dealer

156

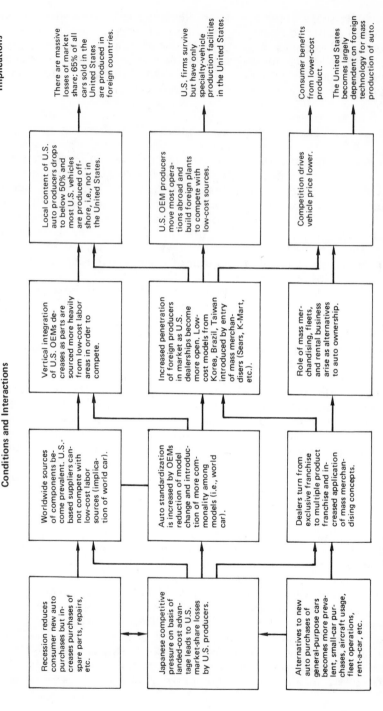

FIGURE 10.2 A maturing industry.

environment, with implications for distribution channels and entry strategies.

The net effect of industry evolution under the maturity scenario is a vastly decreased U.S. auto production base and worldwide competition based in low labor cost countries. One intriguing effect of foreign entry and changes in consumer preferences is a radically different retail environment. U.S. firms survive by sourcing and producing most of their products offshore, while a few specialty segments are served from domestic sources. Under conditions of maturity, plant shutdowns and layoffs would continue, and the United States would likely become dependent on foreign-based technology for mass production of the automobile.

Scenario 3: Fundamental Restructuring

Unlike Scenarios 1 and 2, the third scenario (see Figure 10.3) assumes substantial and continuing pressure for technological innovation. Driven by rising fuel costs, innovation is efficiency oriented and is a major factor in competition. The implications of new product innovations are pervasive. Responding to a rising market demand for both fuel savings and performance, the major U.S. producers seek more innovative technologies from which to develop new product designs. As investment is turned toward new products, the degree of vertical integration declines markedly; specialized, robust, technologically active suppliers become key sources of new technologies. During the period of transition and while new products are developed, U.S. firms lose their market shares in standard models.

Unlike earlier periods, U.S. producers are unable to adequately develop a full-line production capability. Change is rapid, products are diverse, and some firms are financially weakened. As a result, a higher degree of specialization occurs, not only by producers but by country as well. OEMs tend to do the best when they focus on the technology at which they are most expert. The pace of technological innovation quickens, and specialization by technology creates strong domestic linkages; availability of supporting industries (e.g., chemicals, materials) becomes a critical determinant of location.

The transformation of the auto industry from a mature, technologically quiet industry into a hotbed of innovation and change creates opportunities for U.S. firms to attain competitive advantages through development of radically new products. The same, however, can be said of the Japanese and the Europeans. Whether U.S.-based production regains lost market share by creating and exploiting new markets depends on its ability to "out innovate" its competitors.

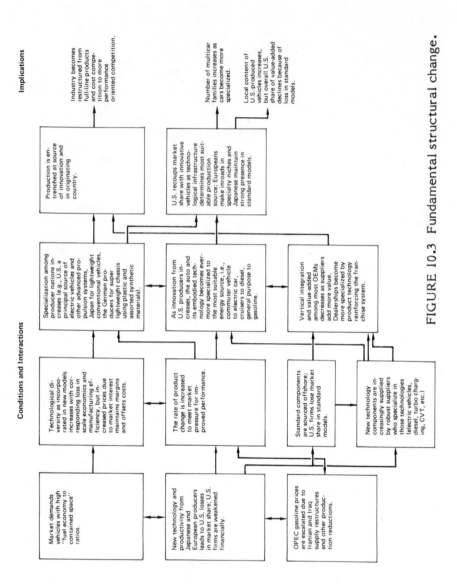

FIGURE 10.3 Fundamental structural change.

In this regard the U.S. firms have historically had an edge on the Japanese, whose advantage seems to lie in refining and manufacturing an established technology. Many of the Europeans, however, have been more performance and technology oriented than the U.S. firms and provide a strong innovative challenge. Thus, Europeans make inroads in the growing specialty segments, while Japanese maintain a strong position in standard models. U.S. firms regain some lost ground through innovation, and, overall, the U.S. share of value-added returns to its 1979-1980 level.

The Scenarios Contrasted

It is clear from our discussion that the future presents significant risks for the domestic auto industry no matter what scenario one picks. Obviously, the outcome in Scenario 3 is more optimistic, but it supposes that U.S. firms can "out innovate" their competitors and (perhaps equally difficult) survive losses of market share and financial difficulties during the period of transition. Prospects are even more bleak in Scenario 2, where the domestic industry is reduced to specialty production; this is effectively the "Britainization" of the U.S. auto industry. The future is not so calamitous under Scenario 1, but even here the domestic industry shrinks and becomes much less profitable than before.

There are two aspects of the industry's environment that we have held constant in developing the scenarios, which may have a bearing were they to change. The first is the overall state of economic activity. We have assumed that the conditions of recession (sluggish, if any, growth; anemic final sales) persist. A return to normal or above-normal rates of growth would be unlikely to affect our conclusions under Scenario 2 (maturity) but may influence developments in Scenarios 1 and 3. The principal effect would be to strengthen the financial position of the domestic firms, allowing them to cover more market segments or develop more new products, thus strengthening their overall market share.

The second critical assumption is the absence of any policy action by the federal government. It is somewhat unrealistic to expect public policy to be on hold over the next several years. Legislation has already been proposed that would impose strong trade restrictions and reduce taxes.[2] Although proponents have made various claims for such proposals, it is important to note that their effects are likely to differ under different patterns of industry development; by meshing the scenarios with alternative policies, their joint implications can be made clear.

IMPLICATIONS FOR PUBLIC POLICY

A full-scale examination of auto industry policy measures is beyond the scope of this study. What we can do is indicate the broad differences in the industry's evolution if two kinds of general policy measures are enacted. The first general category includes "internal" measures, including deregulation, and investment incentives; the second category includes "external" policies to reduce imports. We propose to examine the effects of a representative or generic policy within each category. Within the "internal" category, for example, the types of regulatory measures considered include a freeze on new regulations, postponement of compliance dates, or outright repeal of some regulations; tax incentives would include accelerated depreciation and an increased investment tax credit. On the "external" side, almost all current proposals are labeled "temporary" and are represented by quotas on imports; the basic thrust of such action would be to afford a measure of protection during a temporary transition period.

Before considering the effect of these broad policy options on industry evolution, it is important to note that macroeconomic policy may have an important bearing on the implications of the three scenarios. Events of the last few years have demonstrated the sensitivity of automobile sales to high, real interest rates and sluggish economic growth. Since we assume that the conditions of recession will persist, any developments in macroeconomic policy that are to increase overall economic activity would affect the predicted outcomes as noted above. Likewise, macroeconomic policy that had the effect of deepening the recession, or which created economic instability, could weaken the U.S. firms under each scenario.

The implications of alternative policies are examined under the three scenarios in Table 10.2. The scenarios are listed in the first column, and the three policy regimes are listed across the top; we have added "do nothing" to the "internal" and "external" categories outlined above. The second column establishes a baseline by summarizing the scenarios developed in Figures 10.1, 10.2, and 10.3. The third and fourth columns present likely changes in each scenario if the hypothetical policies were to be enacted.

Perhaps the clearest conclusion to emerge from the analysis in Table 10.2 is that what outcome one predicts for a given policy depends on one's view of the industry's evolution. This is most clearly spelled out in comparisons of the maturing industry scenario and the other scenarios. Under conditions of maturity, neither internal policies nor temporary trade restrictions alter the long-term decline of U.S.-based production. Tax and regulatory changes do strengthen the domestic financial position of firms, and trade measures slow migration of capacity. But unless permanent and restrictive quotas are adopted, the long-term prognosis is little changed.

In the other two scenarios, internal policy options generally strengthen the domestic firms, although the extent of such strengthening is hard to determine without far more specific proposals. Moreover, there are important short-run differences between Scenarios 1 and 3. Under transient misfortunes, the transition period is characterized by increased capital spending, but the assumed parity of new U.S. products implies that market-share losses will not be substantial. In contrast, Scenario 3 envisions a longer transition period in which domestic firms will be under relatively intense pressure because of their cost position.

We have assumed that the changes in internal policies are sufficient to reduce the risk that marginal competitors and product lines will be shut down because of financial weakness. Under both Scenarios 1 and 3, the domestic firms are placed in a somewhat stronger position vis-à-vis foreign competitors, although the migration of standard components is more extensive in Scenario 3. In effect, imports and other market pressures maintain strong incentives for change (i.e., new capacity in Scenario 1; new products, capital, and organization in Scenario 3), while policy changes ease the financial pressure of the transition.

External measures are assumed to have similar effects on the financial side but raise new risks in their impact on competition. We have assumed that a temporary quota raises domestic market shares and prices and thus offers financial support under both Scenarios 1 and 2. The risk is that a reduction of imports will ease competitive discipline and reduce the urgency of change. Take Scenario 3, for example. Without the strong, immediate pressure of import competition, the domestic firms (and their partners) may make the transition in terms of new capital investment and new products but may fail to undertake changes in organization and management that are necessary for long-term viability. It is our view that the temporary nature of the external measures and the strong demands for the market mitigate the competitive cushion offered by trade restriction. We expect Scenario 3 with external policy measure to result in a strong viable domestic industy, somewhat larger than would be the case under no change in policy. Nevertheless, the risk of organizational slack is real and must be weighted in the balance when asessing the options and their impacts.

It is clear from this discussion that predictions about the efficacy of policy must rest on a particular view of the industry's development. Particularly in the case of trade policy, there is no guarantee that temporary action will have its intended effect. If

TABLE 10.2 Industry Scenarios and Government Policy

| | Government Policy Initiatives | | |
Industry Scenarios	No Changes in Policy	Internal (relax regulations; tax incentives)	External (temporary import quotas)
Transient economic misfortune	• U.S. firms lose some market share while making transition to new fleet. • Japanese build plants in the United States and maintain strong position while developing new market niches. • Loss of market share and strong price competition reduce financial strength of U.S. firms. • United States maintains viable domestic production base as U.S. firms regain some market share with new products and as Japanese open new U.S. plants; United States loses preeminent position in production and technology.	• Tax incentives spur capital investment and ease financial pressure; U.S. firms lose market share in transition, but new models and new facilities come on stream somewhat faster. • Added financial strength allows U.S. firms to compete in additional segments; some inroads by Japanese and European producers are cut back. • Japanese maintain strong U.S. market pressure, but pressure on European markets is increased. • U.S. firms maintain viable domestic industry with somewhat larger U.S. firm component.	• Quota umbrella eases financial pressure by raising domestic market share and car prices. • Reduced import presence lowers pace of competition during transition and risks slackened drive for changes in workforce management and organization. • Japanese accelerate building of plants in the United States. • New capacity is added, and firms compete in new segments with additional financial resources; expiration of quotas is followed by wave of new Japanese and European products.
A maturing industry	• Extensive migration of capacity to low-cost offshore locations (OEM and suppliers). • Increased penetration of products from low-cost sources; massive loss of domestic market share—65 percent of all cars sold in United States are produced in foreign countries.	• Migration of capacity and imports of standard models continue unabated by tax incentives. • Capital investment in specialty and high-technology models is spurred by incentives and regulatory reform. • Mass production migrates	• Migration of capital is slowed, but protection encourages production inefficiencies and high prices, weakening primary demand. • Cost gap between U.S.- and foreign-based production grows, thus increasing pressure for offshore operations.

163

- U.S. firms survive with offshore production of basic models but are weakened financially; only specialty products are made in the U.S. firms.
- United States becomes dependent on foreign-based technology for mass production of autos.

offshore; the United States emphasizes and retains the capability for high technology and specialty autos and associated small-lot, high-cost production.

- Unless made permanent, import restrictions have no long-term effect on maturity scenario.

Fundamental structural change

- U.S. firms lose market share in standard models and are weakened financially; some firms are in marginal position.
- Standard components are sourced offshore.
- Increasing real price of oil creates incentive for rapid and radical innovation; technologies are specialized by use and country.
- Transition period is lengthy; U.S. firms that survive transition recoup some market-share losses with innovative products; Japanese remain strong in standard models, and European make inroads in specialty niches.
- Local content of U.S.-produced vehicles rises but, overall, U.S. share of value-added declines because of loss of standard models.

- Incentives and reforms ease financial pressure, but U.S. firms lose market share in standard models.
- Tax incentives and deregulation aid investment in product development and capacity improvement for innovative components.
- The migration of capacity for traditional components is extensive; wage differentials are steep compared with tax incentives.
- U.S. firms survive in somewhat stronger position with innovative products, but U.S. share of value-added is lower because transition losses in standard models are not fully recouped. Japanese and European producers remain strong.

- Technology ferment imparts qualities of an infant industry.
- Protection umbrella eases financial pressure and preserves market share temporarily in standard models.
- U.S. firms weather transition in stronger financial condition, but protectionism increases risk of reducing urgency of structural changes; strong market pressures for innovation continue.
- United States regains strong market position with innovative products; its share of value-added is little changed; Japanese imports and European products reemerge after protection is lifted.

the maturity scenario prevails, only long-term measures will preserve a large domestic industry. Yet permanent policies would not be neutral in their effects on adaptation under Scenarios 1 and 3. Especially in the case of Scenario 3, a policy of permanent quotas, essential to retention of domestic production under maturity, would likely slow the industry's adjustment, both in organization and in technology. The result could well be a weaker industry in terms of performance and technical leadership than would be the case with temporary trade measures.

IMPLICATIONS FOR MANAGEMENT: IN THE STARS OR IN THEMSELVES?

The policy options examined in this section may have an important bearing on the evolution of the industry, depending on what scenario or mix of scenarios turns out to be true. We have seen that where change and adaptation are required, public policy can increase the industry's flexibility and freedom of action. While public policy thus has a critical supporting role to play, a role likely to be focused on mitigating risks and facilitating necessary changes during the transition period, the long-term future of the industry is by and large in the hands of its participants. This is not to minimize the challenges they face nor the impacts of external factors. It is only to reaffirm the central role of the productive confederation--the firms, the unions, the suppliers--in charting the industry's course.

This report identified a number of implications for management in the auto industry, implications that hold out the prospect of substantial changes in the way the business is managed. Some of our conclusions are generally applicable, while some depend on a particular scenario. Under "maturity" for example, the key to competition is the ability to manage a worldwide production and distribution system, with worldwide sourcing and technical innovation that extends and refines existing concepts. In the case of a "fundamental change," however, the competitive tasks are twofold: (1) improvements in quality and productivity in existing models and (2) the development and introduction of new concepts in products and processes. These two scenarios may thus require different organizational and managerial capabilities.

Whatever the specific path of development may look like, there are several implications for management that appear to be generally applicable. While the evidence in this report suggests some room for improvement in automation and new-process technology, the bulk of the productivity-quality gap lies in established manufacturing practices (including design and production), methods of organization, manufacturing systems, and the management of workers. These aspects of the firms are as much a capital resource as buildings and equipment. As we noted in Chapter 7 in our discussion of workforce management, change requires investment, not only in procedures or structure but also in habits, relationships, and even basic concepts of the business. It is the investments in such activities rather than the more publicized investment in new plants that hold the key to regaining competitive parity with the Japanese.

The thrust of these investments in capital, people, and organization is more than a change in the product mix--it is to achieve superior manufacturing performance, to make manufacturing a major competitive factor. We have examined changes in the management of people, the need for a more open agenda between firms and employees, and a move to engage the work force in the competitive activities of the enterprise. But other equally fundamental changes also may be required. At the same time that the domestic firms are faced with a competitive challenge in the cost and quality of established products, the importance, even the necessity, of significant product innovation may grow over the next few years. If so, competitive advantage in the mid- to late 1980s and 1990s will depend on the ability to develop and implement new design concepts in a relatively uncertain environment.

Competition in these terms would confront the existing domestic producers with the need for different organizational capabilities than those that have been developed over the past 40 years. In an environment where products were standardized and innovations incremental, the successful firms excelled in exploiting economies of scale, in incremental innovation, and in coordination and control. Entrepreneurship and brilliant but risky projects became increasingly dysfunctional, while the structure of organization became more hierarchical and its processes more bureaucratic.

Contrast this state of affairs with an organization that excels at innovation and at coping with uncertainty. It is likely to be guided by individuals willing to take risks when precise calculations cannot be made and to utilize organizational structures that are highly interactive and adaptive, where negative feedback is likely to be acted upon promptly.[3] This is, of course, an ideal type, and there is as yet no reason to suppose that the successful auto producers will function like hi-tech electronics firms. But achieving radical innovation would seem to require an organizational setup different from that appropriate for efficient production of an established design.

One need only look at Ford in the 1920s to see that technology-based competition may require significant innovation in organization and management as well as in products and processes. As an example of the kind of changes required, consider the role of research. In a period of technical ferment, new ideas and new concepts are vital to competitive success. Yet for the most part the research organizations in the auto industry have not been tied in to the basic competitive activities of the business, simply because innovation that required research was not essential to competitive success. An R&D organization that operates in its dominate mode as an applied engineering group tends to relegate basic or even applied research to the back seat; it is likely to be different from the organization one would design to generate new, competitively significant and viable concepts. Not only would the lines of communication and the reporting relationships be different, but changes would be likely in the types of people employed, the process for project selection, funding, and so forth.

Other examples could be cited. The point is that the organization, the people, the concepts that are likely to be essential to competitive success in the auto industry in the 1980s and 1990s are different from those that have prevailed in the postwar period and different from those that prevail today. We have emphasized the magnitude of the challenge but have not underscored the strengths in the industry or the "window of opportunity" that the current crisis has opened. The critical element in the industry's efforts to tap resources and exploit its considerable strengths appears to be strategic vision, the ability to see the future, to see the business in a way that is different from the way it has been seen in the past. Given the central importance of change and adaptation to the industry's future, it is perhaps fitting to conclude this report with Alfred Sloan's commentary on Henry Ford, long after General Motors had established its dominance in the marketplace.

> Mr. Ford's concept of the American market did not fit the realities after 1923. [He] failed to realize that it was not necessary for new cars to meet the need for basic transportation. . . . Mr. Ford, who had had so many brilliant insights in earlier years seemed never to understand how completely the market had changed from the one in which he had made his name. . . . The old master failed to master change.[4]

NOTES

1. It is important to remember that our perspective is a national industry perspective. We do not attempt to judge the impact on specific producers, nor are we explicitly concerned with the future of U.S.-based companies. Our focus is on U.S.-

based <u>production</u> and the U.S. share of value-added in the industry.

2. The recent "voluntary" restrictions adopted by the government of Japan establish a short-term barrier to further penetration; the trade issue may well be raised once they are lifted.

3. This characterization of the dynamic enterprise is based on work by Burton Klein. See Klein (1977) for a full statement of his views.

4. Sloan (1972), pp. 186-187.

Estimates of Comparative Productivity and Costs Under Alternative Methods

This appendix presents estimates of the U.S.-Japanese landed-cost differential on a small vehicle, using several different methods and approaches. In addition, estimates of differences in productivity also were obtained. Some of the panel members have had access to internal studies of these issues that make use of proprietary data. Our intent here, however, is to make use of public information in order to illustrate the range of estimates implicit in generally available data.

The four approaches can be distinguished in their unit of analyses. The first adopts what is essentially an economy-wide macro perspective; this we have labeled the industry/macro approach. The second, the industry/micro view, looks at the issue from the standpoint of the industry taken as a whole. The third, the company perspective, uses data on two major firms, as detailed in their annual reports. Finally, we present data on plant-by-plant comparisons of particular kinds of production processes. The concluding section of this appendix summarizes the results.

Before reviewing the analysis it is important to note the difficulties associated with calculations of this sort. The auto industries of the United States and Japan produce a different mix of products and have organized production in different ways, particularly in terms of vertical integration. Productivity comparisons are also significantly affected by differences in capacity utilization that have been substantial in recent years. While attempts have been made to correct for these factors, even the most careful comparison requires judgements and assumptions that affect the results.

The Industry/Macro Approach

One way to compare productivity in auto manufacturing in the United States and Japan is to do the comparison on an economy-wide basis. This approach, which involves comparisons of efficiency in every production activity required in making an automobile (i.e., mining iron ore, oil refining, steelmaking, machinery fabrication, power generation, business services), uses an input-output analysis. Given the scope of automobile production, this is close to comparing the economy-wide productivity of the United States with that of Japan.

Available data on economy-wide productivity indicate a U.S. advantage, so it is no surprise that the industry/macro approach tends to find higher total labor hours per vehicle (defined over the whole economy) in Japan. Dan Luria of the United Auto Workers' Research Staff has used 1977 input-output data updated to 1980 and estimated labor content for a small, 1980 Japanese vehicle to be 364 hours versus 336 for the American; this is a U.S. advantage of about 8 percent.[1] With assumed hourly charges of $11 (averaged over all embodied hours) for the United States and $7 for Japan, Luria's estimate of the manufacturing cost advantage of the Japanese is $1148. If freight charges ($400) are subtracted, Luria's analysis implies a landed-cost advantage of $748.

It should be noted that, while the industry/macro analysis does not provide an estimate of productivity or cost difference originating within the auto sector, it is not inconsistent with the existence of a U.S. disadvantage. Indeed, existing evidence suggests that a good part of the production activity outside of the auto sector (and even some in the sector but outside of the big firms) may take place in small establishments where productivity is low. Luria suggests that the Japanese had a productivity advantage of 11 percent (about 15 percent in 1981 terms) in the auto sector but that they were 16 percent less productive in other industries.

Industry/Micro Costs and Productivity:
Motor Vehicle and Parts Industry

To obtain estimates of cost differences in the auto industry itself, it is useful to first examine differences in labor productivity. In 1974 Baranson estimated that output per labor hour in the Japanese motor vehicles and parts industry was 88 percent of the level reached in the United States (i.e., the ratio of productivity in Japan to productivity in the United States was 0.88).[2] This 12 percent U.S. advantage in 1974 is consistent with calculations developed by the British Central Policy Review Staff; using 1973 data they estimated the relative productivity ratio to be 0.82.[3] These estimates can be updated using published data on growth rates of productivity in the United States and Japan. Abstracting from cyclical fluctuations, the evidence suggests that growth in labor productivity in the Japanese auto industry (vehicles and parts) averaged 8-9 percent per year in the 1970s; the comparable figure for the United States is 3-4 percent.[4] If we apply a mid-range estimate of the differential (i.e., 5 percent) to previous estimates of relative productivity, we arrive at a value for 1980 of 1.18. The data on industry growth rates imply that the Japanese producers operated at levels of productivity almost 20 percent above their American competitors.

More rapid growth in productivity in Japan has been accompanied by higher rates of wage increase. Whereas in 1974 Japanese hourly compensation rates were about 37 percent of the U.S. figure, in 1980 the ratio was roughly 50 percent.[5] Relative unit labor costs can be calculated by dividing the compensation ratio by the index of relative productivity.[6] This method yields an estimated unit labor cost ratio of 0.424 = (0.5/1.18) for 1980. Thus, Japanese-U.S. differences in the growth of productivity and compensation have been offsetting; the wage gap has continued to narrow, while a productivity gap has emerged and grown larger; relative unit labor costs remained roughly constant at 0.425 over the 1974-1980 period.

We assume that the estimate of relative unit labor costs applies both to the auto manufacturers and to their suppliers. The estimate can be used to calculate the dollar value of the Japanese advantage in lower overall labor costs per vehicle. Since our focus is on the overall labor content in the vehicle, including purchased parts and materials, we are in effect comparing the relative position of the Japanese and U.S. automobile production systems. We are not comparing labor costs at the level of a Toyota or a General Motors' (GM) car; our analysis seeks to estimate the impact of productivity and compensation in the whole productive confederation--original equipment manufacturers (hereafter, OEMs) and suppliers of components and materials. Note that this is different from the industry/macro approach, which included many more activities outside the "productive confederation" in the calculations.

Table A.1 presents the basic building blocks of the analysis. Column 1 contains an estimate of the shares in manufacturing costs of hourly and salaried labor (at the OEM level), purchased components, and materials. These estimates are based on data prepared for the National Research Council, Committee on Motor Vehicle Emissions, as well as discussions with industry sources.[7] The estimates do not reflect the experience of any one company but are intended to approximate an industry average. It should be emphasized that all of the data refer to production of a small, subcompact vehicle. In addition to the data in column 1, industry participants have provided us with estimates of average OEM labor hours per vehicle, current rates of employee cost per hour,

172

TABLE A.1 Calculation of U.S. and Japanese Labor Costs for Subcompact Vehicle[a]

Cost Category	Share in OEM Manufacturing Cost (U.S.) (1)	Average Hours Per Vehicle (U.S.) (2)	Estimated OEM Employee Cost Per Hour (dollars) (U.S.) (3)	Estimated Cost Per Vehicle (dollars) (U.S.) (4)	Labor Content (percentage) (U.S.) (5)	Labor Cost Per Vehicle (dollars) [(4) × (5)] (U.S.) (6)	U.S.-Japan Difference (dollars) [(6) × 0.575] (7)
Hourly OEM labor[b]	0.24	65	18.00	1170	100	1170	673
Salary	0.07	15	23.00	345	100	345	198
Purchased components	0.39	N/A	N/A	1901	66	1255	721
Purchased materials	0.14	N/A	N/A	683	25	171	98
Total	—	—	—	4875[c]	N/A	2941	1690

NOTE: Totals may not add because of rounding.

[a] Calculations assume an exchange rate of 218 yen per dollar.
[b] OEM hourly labor is defined as total nonexempt and includes direct and indirect production workers.
[c] This figure is total manufacturing cost and includes labor, materials, and manufacturing overhead of 16 percent.

SOURCE: Committee of Motor Vehicle Emissions (1974); staff reports of U.S. companies; and memoranda from panel members.

and labor content in purchased components and materials; these data are presented in columns 2, 3, and 5.[8]

The calculation of U.S.-Japanese cost differences takes place in three steps. We first use the data in columns 2 and 3 to get an OEM hourly labor cost per vehicle of $1170, and then extrapolate using the cost shares (column 1) to arrive at a total manufactured cost and the cost of purchased components and materials (column 4). Next, we multiply the cost per vehicle in column 4 by an estimate of the labor content of the three categories presented in column 5. The data imply, for example, that $1255 of the $1901 cost of components is labor cost. Finally, we calculate the U.S.-Japanese labor cost gap by multiplying the U.S. data in column 6 by 0.575; the adjustment factor is based on our previous estimate of the Japanese-to-U.S. unit labor cost ratio.[9] Thus, column 7 provides an estimate of the difference in the cost of producing a small vehicle in the United States and in Japan due to differences in unit labor costs, not only at the OEM level but also at the supplier level as well.

The estimated cost gap is sizeable. When the effects of components and materials suppliers are added, the Japanese cost advantage is $1690. Although the calculations in Table A.1 are based on estimates of cost structure and labor content, reasonable adjustments of these assumptions would not reduce the order of magnitude of the Japanese cost advantage. To arrive at a landed (i.e., after shipment to the United States) cost differential, it is necessary to add general administrative and selling expenses as well as the costs of capital and transportation. Our estimates from annual reports and other sources suggest that these factors would reduce the Japanese advantage to $1436.[10]

The Company Perspective: Evidence from Annual Reports

Additional insight into the differences in production costs in the United States and Japan can be obtained through an analysis of data contained in company annual reports. The use of annual reports shifts the focus of analysis to the costs incurred by the major manufacturers. In terms of labor costs the shift in focus generally means that no information will be available on labor embodied in components or materials. However, differences in costs associated with nonlabor inputs and with corporate-wide management and salaried personnel can be assessed. Furthermore, the annual reports allow us to estimate labor productivity at the OEM level. This approach thus provides a useful check on industry estimates.

Any comparison of Japanese and U.S. companies must confront several analytical problems. Perhaps the most serious issue is the great difference in vertical integration and relationships with suppliers. At Toyota, for example, purchases account for almost 80 percent of the value of final sales. Because Toyota holds an equity interest in many of its suppliers, this figure is somewhat misleading. Comparable data for U.S. firms show much less reliance on suppliers. GM, for example, has a purchase-to-sales ratio of less than 50 percent. A second major problem is the different product mix of U.S. and Japanese firms. The data we shall use are those for 1979, when the products of the Big 3 were dominated by models in the medium-size ranges. The Japanese produced a much narrower range of vehicles, with heavy emphasis on the subcompact segment.

Our approach to annual-report analysis can be illustrated using data on Toyo Kogyo (Mazda) and Ford. Both companies provide sufficient information on automotive production and employment to permit calculation of labor hours per vehicle in U.S. and Japanese operations. Although Toyo Kogyo is about one-third the size of Toyota and used to be a relatively high-cost producer, it has experienced significant gains in productivity in recent years and now appears to have costs that are on par with those at Toyota and Nissan. Available evidence suggests that Ford is somewhat less efficient than GM, so that the Ford data may understate industry productivity.[11]

The basic estimates of employee costs per vehicle are presented in Table A.2. Two principal assumptions underly the calculations. Data on total domestic employment and total domestic employee costs were broken down into automotive and nonautomotive components based on the ratio of automotive to total sales. This effectively assumes that nonautomotive businesses were as productive and as labor intensive as the automotive group. Since cars and trucks account for over 90 percent of sales at Ford and Toyo Kogyo, this assumption is not critical. The second assumption is that Ford employees worked an average of 1620 hours per year. For Toyo Kogyo the comparable number was assumed to be 1900 hours. These adjustments reflect differences in the effects of vacations, holidays, personal leave, and absenteeism.[12]

The evidence in Table A.2 reveals sizeable differences in productivity and total employee cost per unit. Given our assumptions, we estimate that the average Ford vehicle required 112.5 employee hours, while Toyo Kogyo produced an average vehicle in only 47. At an exchange rate of 218 yen to the dollar, we find that employee cost in the average Toyo Kogyo vehicle was less than $500; the comparable figure for Ford was $2464. The sizeable cost gap reflects differences in product mix and vertical

TABLE A.2   Estimated Employee Costs Per Vehicle, 1979

	Ford	Toyo Kogyo
(1) Domestic production of cars and trucks (millions)	3.163	0.983
(2) Total domestic employment		
Automotive	219,599	24,318
Nonautomotive	19,876	2,490
(3) Total domestic employee hours		
Automotive (millions)	355.75	46.20
(4) Total employee cost		
Automotive (millions)	$7794.50	$482.20
(5) Employee hours per vehicle	112.5	47.0
(6) Employee cost per vehicle	$2464	$491

NOTES: Line (1): Published production figures for Ford have been adjusted to eliminate 65,000 imported vehicles; the Toyo Kogyo data have been adjusted for production of knock-down assembly kits. Lines (2)-(4): Data on automotive employment and costs were obtained by assuming that the ratio of automotive employment to total employment was the same as the ratio of sales; the same assumption was made to obtain Ford employment costs. Line (3): Ford hours were determined by assuming that each employee actually worked 1620 hours per year. Toyo Kogyo hours assume that each employee actually worked 1900 hours. Line (4): Data include salaries, wages, and fringe benefits. For Toyo Kogyo, compensation data were derived by updating a 1976 figure using compensation growth rates at Toyota; an exchange rate of 218 yen per dollar (1979 average) was used to convert yen to dollars.

integration as well as wage and productivity differentials. Toyo Kogyo concentrates heavily on the production of small cars, while Ford's product line covers a much wider range of sizes. Ford produces a larger fraction of the average vehicle in house. Information on value-added in the annual reports and discussions with industry sources suggests that the Toyo Kogyo results should be increased by 15-20 percent in order to adjust for differences in vertical integration. Using the higher estimates yields 56 hours per vehicle for Toyo Kogyo.[13]

To correct for differences in mix we have estimated the cost to Ford of producing the Toyo Kogyo product mix. The calculations are presented in Table A.3. The procedure uses data on manufacturing costs by vehicle size class developed for the Committee on Motor Vehicle Emissions of the National Research Council in 1974.[14] Estimates of the cost to Ford of producing the Toyo Kogyo mix were obtained by first computing a weighted average of the relative manufacturing cost indexes with Ford's 1979 production shares by size as weights. The ratio of the comparable Toyo Kogyo weighted average (1.06) to the Ford weighted

TABLE A.3   Product Mix Adjustment

	Ford	Toyo Kogyo
(1) Ratio of car to total vehicle production	0.645	0.652
(2) Production shares by size		
small	0.11[a]	0.83
medium	0.68[a]	0.17
large	0.21	—
(3) Relative manufacturing cost by size (small = 1.00)		
small	1.00	N/A
medium	1.35	N/A
large	1.71	N/A
(4) Weighted average relative manufacturing cost (small = 1.00)	1.38	1.06
(5) Production of Toyo Kogyo mix at Ford level of integration		
(a) employee cost per vehicle	$1893[b]	$589
(b) employee hours per vehicle	87[b]	56

[a] Assumes that only Pinto and Bobcat models are small; Mustang and Capri sales were placed in the medium category.

[b] Obtained by multiplying lines (5) and (6) in Table A.2 by (1.06 ÷ 1.38).

SOURCE: Committee on Motor Vehicle Emissions (1974); *Ward's Automotive Year Book;* annual reports.

average (1.38) was used to adjust both costs and productivity. It is an estimate of the effect of product mix on Ford's average cost and labor hours per vehicle. After these adjustments we estimate that Ford would require 87 employee hours to produce the average-size vehicle in the Toyo Kogyo product line, compared to 56 hours in the Japanese firm. Labor cost per vehicle is just over $1300 higher at Ford. These comparisons are based on the average-size vehicle at Toyo Kogyo. For a small vehicle (i.e., Pinto versus Mazda GLC) the Ford estimate is 82 hours per vehicle, while the comparable Toyo Kogyo figure is 53; the corresponding costs per vehicle are $1785 (Ford) and $556 (Toyo Kogyo). Even this adjustment may overstate costs and hours required to produce the Toyo Kogyo mix at Ford if the trucks and commercial vehicles produced by the two companies differ substantially.

The analysis of annual report data suggests that the difference between Ford and Toyo Kogyo employee cost per small vehicle in 1979 was about $1200. Updating to 1980 would increase the absolute dollar difference by about 10 percent, a reflection of changes in wage rates and materials prices. An adjustment for changes in exchange rates also would have only negligible effects. We have used an exchange rate of 218 yen per dollar (1979) average; use of 200 yen per dollar (approximate rate at the end of 1980) would reduce the gap by about $50.[15]

The estimated cost differential reflects the compensation and productivity of all employees in the two firms. It does not, however, capture differences in unit labor costs in components or materials. Although the data were developed within a different framework, the evidence on labor content in components and materials used earlier is suggestive of the likely order of magnitude. After adjusting for possible differences in productivity differentials at the supplier level, adding $700 to the employee cost differential seems justified.[16] We conclude that differences in compensation and productivity lead to significantly lower unit employee costs in Japan; if Ford is indicative of average U.S. performance and if Toyo Kogyo is representative of the major Japanese auto producers (and comparisons with Nissan and Toyota suggest it is), then the cost advantage from labor and materials is likely to be about $1900 per vehicle. Analysis of other elements of total cost--e.g., selling and general administrative expenses--provides evidence of a U.S. advantage of about $135 per vehicle. Likewise, freight costs and the 2.9 percent tariff add another $400 to Japanese costs. The net result is a landed-cost advantage to the Japanese producers of $1465.

Evidence from the annual reports suggests a somewhat larger cost differential than we obtained using the industry/micro approach. The difference arises primarily from a much higher level of productivity in Toyo Kogyo than suggested by our updating of the 1974 estimates of Baranson. Those estimates applied to the entire motor vehicles and parts industry and were admittedly rough and imprecise. It is not unreasonable that relative productivity at the OEM level would exceed levels achieved by suppliers. Without additional evidence it is difficult to judge the accuracy of the Toyo Kogyo data, yet similar calculations for the other major producers and discussions with industry experts suggest that the Toyo Kogyo analysis is representative. Indeed, care has been taken to make sure that our assumptions erred in the direction of underestimating the gaps in productivity and cost.

Productivity at the Plant Level

The final analytical approach involves a comparison of productivity and cost on a plant-by-plant basis. The sources of these data are government surveys, plant visits by executives and engineers from the U.S. firms, and consultant reports.[17] Care has been taken in these studies to compare similar processes and to correct for differences in product mix and degree of vertical integration. There is no claim made of complete coverage; only a few types of processes have been examined. It is felt, however, that several of the most critical elements in automobile production have been studied and that those studied are representative of the industry average.

Table A.4 presents data on labor hours per vehicle in selected plants in the United States and Japan. It is evident that the Japanese have a sizeable overall productivity advantage and that the differential varies considerably in parts of the process. In transmissions, foundry, and forge operations the differential ranges from 0 to 35 percent. The largest gap is in the stamping plants, where the Japanese advantage is close to 3 to 1. Stamping is one of the few processes where the Japanese appear to have a significant technology edge. The major press lines and transfer presses in Japan are equipped with U.S.-made automatic rolling bolsters and quick-die-change features.[18] Together with other automatic devices these features allow the Japanese to achieve output rates of 550 panels per hour versus 300-325 in the United States.

The plant comparisons do not provide a complete set of cost figures, but the data can be used with previous information to obtain estimates of employee cost per vehicle. The comparisons imply that the ratio of Japanese to U.S. productivity at the OEM level is 1.9. Using an employee cost per hour ratio of 0.5, the productivity evidence implies a unit labor cost ratio of 0.263. If we apply this figure to our previous estimate of U.S. OEM employee costs per small vehicle of $1515, we end up with a

TABLE A.4   U.S.-Japanese Difference in Productivity in Selected Plants: Hours per Vehicle

Country	Plant/Process							
	Assembly	Stamping	Engine	Transmission	Axle	Foundry	Forge	Total
United States[a]	38	10	7	8	5	5	1	74
Japan[b]	17	4	4	6	3	4	1	39
Difference	21	6	3	2	2	1	1	35

[a] SOURCE: J. E. Harbour, *Comparison and Analysis of Manufacturing Productivity* (final consultant report), Harbour and Associates, Dearborn Heights, Mich., 1980, p. 2.
[b] SOURCE: Japanese Ministry of Labor, *Statistical Survey of Labor Productivity,* 1978, as cited in Abernathy *et al.* (1980, p. 41).

Japanese advantage of $1094. This assumes, of course, that the higher estimated productivity ratio is reflected only in lower Japanese hours, not in higher (and more expensive) U.S. hours. The procedure may thus understate the cost gap.

The Impact of Capital

The difference in labor productivity estimated here may be affected by differences in capital, which may in turn affect cost comparisons.

Before examining evidence on this point it may be useful to clarify the issues with a simple diagram. Figure A.1 (panel 1) presents two unit isoquants, one for the United States (Q_{us}) and one for Japan (Q_j). The vertical axis measures capital, and the horizontal measures labor. The isoquant depicts all the possible combinations of capital and labor that can be used to produce one unit of output. The way we have drawn the diagram implies that at a given capital-labor ratio the Japanese use less capital and labor to produce a unit of output than U.S. firms. This may be due to differences in management or techniques.

We assume that the United States is at point A, with a capital-labor ratio given by the slope of the ray, OA. The Japanese are at point B, with a higher capital-labor ratio. At these points the Japanese have higher labor productivity and lower capital productivity. Assume for the moment that the unit isoquant for Japan in Figure A.1, panel 1, shifted upward to coincide with the unit isoquant for the United States, while Japan's capital-labor ratio remained unchanged. Japan would be at point C, while the United States is assumed to remain at point A. Compared with the United States, labor productivity in Japan is much higher, but capital productivity is lower. Depending on the prices of the inputs, total cost comparisons could go either way.

Panel 2 uses the same type of diagram to illustrate a Japanese productivity advantage in both capital and labor. Here Q_j is placed far enough below Q_{us} that both less K and less L are required per unit, even though the K-L ratio is higher.

To examine these issues we have developed estimates of capital-labor ratios and capital productivity using data from Ford and Toyo Kogyo for 1979. The estimates can therefore be compared to the productivity analysis for the two companies presented earlier. The results of our calculations are presented in Table A.5.

Any attempt to compare capital stocks in two countries must confront problems of inflation and differences in prices and currency values. Both problems are addressed in Table A.5, but we also present unadjusted values for comparison. The annual

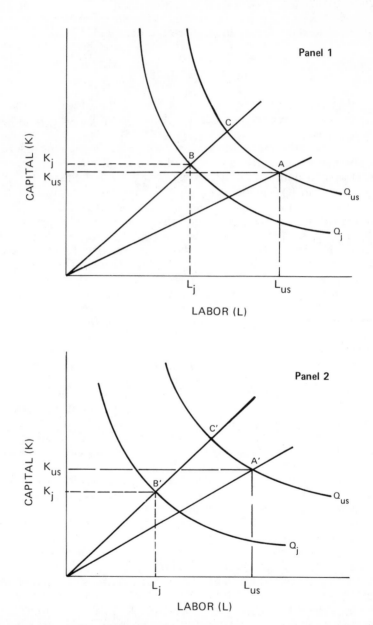

FIGURE A.1 Capital-labor ratios: United States and Japan.

TABLE A.5   Capital-Labor Ratios and Capital Productivity, 1979

Category	Ford	Toyo Kogyo	
		General Exchange Rate[a]	Specific Exchange Rate[b]
Plant and Equipment (billions of dollars)			
Book value	$15.330	$1.539	$1.269
Adjusted for inflation[c]	26.975	2.509	2.068
Inventory (materials and work in process—billions of dollars)	$2.939	$0.077	$0.077
Capital-Labor Ratios (dollars) Capital per employee[d]			
Book value	$40,063	$ 63,113	$52,003
Adjusted	65,599	100,888	83,128
Capital per labor hours worked[e]	$24.73	$33.22	$27.37
Adjusted	40.49	53.10	43.75
Capital Productivity (dollars) Capital per vehicle[f]			
Book value	$3,048	$1,639	$1,351
Adjusted	5,052	2,623	2,161

[a] Average 1979 exchange rate was 220 yen per dollar.

[b] Capital information exchange rate (structures and equipment) calculated based on data in Irving B. Kravis *et al., International Comparisons of Real Product and Purchasing Power* (Baltimore, Md.: The Johns Hopkins Press, 1978); 267 yen per dollar in 1979—applies only to gross plant and equipment.

[c] To correct plant and equipment for inflation, book values were restated in 1979 prices based on age of capital and variation in equipment and structures deflator from national income accounts; only estimated capital used in vehicle production is included (calculated based on fraction of total sales in automotive).

[d] Capital includes both gross plant and equipment and inventories; no inflation adjustment is made for inventories.

[e] Annual hours per employee are 1620 for Ford and 1900 for Toyo Kogyo.

[f] Vehicles include cars and trucks; excludes tractors for Ford.

SOURCE: Annual reports, 1979. Deflators were taken from "Monthly Finance Review," Japan Ministry of Finance (Dec. 1980); "Quarterly National Accounts Bulletin," OECD; "Historical Statistics" OECD; *Economic Report of the President,* 1981, Council of Economic Advisors. Purchasing power parity exchange rates were taken from Irving B. Kravis *et al., International Comparisons of Real Product and Purchasing Power* (Baltimore, Md.: The John Hopkins Press, 1978).

reports for 1979 provide data on the gross book value of plants (including land) and equipment and the value of work in-process inventories. The yen values for Toyo Kogyo have been translated into dollar values in two ways. The first uses the general exchange rate of 220 yen per dollar, while the second uses a specific or purchasing power parity exchange rate of 267 yen per dollar. This latter measure is derived by comparing the yen and dollar prices of comparable (in terms of quality, performance, function, etc.) equipment and structures at a point in time. We have used the estimates prepared by Kravis et al. for capital formation (structure and equipment) in 1973, updated by differences in the rates of inflation in the equipment and structures deflators from the United States and Japan national income accounts. The results for book value show Ford with 15.3 billion dollars, and Toyo Kogyo with 1.54 billion dollars or 1.27 billion dollars depending on the exchange rate used.

Inflation affects comparisons because of the accounting convention of reporting assets at original cost. We have restated the book values in 1979 dollars using rough estimates of the age of the capital stock in the two companies and the business fixed investment deflators from the national accounts. The formula can be written as follows:

$$GBV_{adj} = GBV \left[\frac{P(1979)}{P(1979 - Age)} \right] ,$$

where P is the investment deflator, the number in parenthesis is the year in which the price index is measured, and age is calculated as the ratio of accumulated depreciation to annual depreciation. Since the ages of the capital stock at Ford (8.1 years) and Toyo Kogyo (9.4 years) are comparable, and since the rates of inflation have not been greatly different, the inflation adjustment has only a small effect on the relative amounts of plant and equipment. The data on age of capital suggest that U.S. and Japanese auto firms are using equipment of comparable vintage. It does appear, however, that the Toyo Kogyo data may overstate the average age of the capital stock for all Japanese automobile manufacturers. Data on Nissan, for example, show an average age of capital of 7.1 years in 1979. Likewise, the Ford data may understate the average age of the capital stock for U.S. automobile manufacturers. Data on GM, for example, show an average age of capital of 11.4 years in 1979. These data imply that the capital stock in the U.S. auto industry is, on average, 3-4 years older than the capital stock in Japan. For comparisons made in 1979 this implies that the average piece of equipment in Japan was purchased in 1972, while the average piece of equipment in the U.S. firms was purchased in 1968-1969.

If there were significant technological breakthroughs between 1968 and 1972, the difference in age of capital between the U.S. and Japanese auto industries could imply important differences in the level of technology. Research by Abernathy on the development of the production process in the auto industry suggests that technology in the late 1960s and early 1970s was not characterized by major changes.[19] While a difference of 3-4 years in the age of the capital stock is likely to be of some importance in productivity comparisons, the implication of existing evidence on the age of capital and the course of technical change in the auto industry is that the level of technology embodied in plant and equipment in the U.S. auto industry is not greatly different from the level found in the auto industry in Japan. However, more detailed research on the nature and timing of technological change in the auto industry is needed before strong conclusions can be drawn.

After adding material and work in-process inventory (which is not inflation adjusted) to gross plant and equipment, line III presents measures of capital-labor ratios. We first calculate capital per employee and then take differences in annual hours worked into account. The data support the conclusion that Toyo Kogyo operates with higher capital-labor ratios, irrespective of the measure of labor input used. The estimates of the Toyo Kogyo edge range from 58 percent for the unadjusted book values per employee to 8 percent after correcting for inflation, purchasing power parity, and differences in annual hours worked. These conclusions are unchanged if data for 1978 or 1980 are used instead, implying that adjustment for utilization differences does not affect the results.

The important point to note is that the differences in capital-labor ratios, even without any adjustments, are not sufficient to explain the productivity gap observed earlier. As far as differences in OEM hours per vehicle are concerned, the effect of capital depends on capital's share in the cost of value-added. Even if this were as high as 0.5 (for the economy as a whole it is more like 0.25-0.30), capital differences might explain a labor productivity difference of 30 percent, rather than the difference of 55 percent we actually observe. If we use a capital share of 0.3, the potential impact of differences in capital-labor ratios between the United States and Japan on labor productivity ranges from 2.4 to 17.3 percent, depending on the specific definition of capital and labor used.

The evidence in line III suggests that differences in capital-labor ratios cannot fully explain the productivity gap. It appears that more is involved than substitution of capital for labor. Line IV underscores this point by comparing capital per vehicle. As they stand, the estimates suggest that panel 2 of Figure A.1 is a

TABLE A.6   Summary of U.S.-Japanese Cost Differences

Cost Category (per vehicle)	Approaches			
	Industry/Macro (1)	Industry/Micro (2)	Company (3)	Plant (4)
OEM employee cost	N/A[a]	$871	$1219	$1115
Materials and components	N/A[a]	$819	$700[b]	$700[b]
Other costs	N/A[a]	$146[c]	$146[c]	$146[c]
Freight and duties	N/A[a]	($400)[d]	($400)[d]	($400)[d]
Landed-cost difference	$748	$1436	$1665	$1561

[a] These data are not included in the analysis; sources used only provided a total cost difference.

[b] This estimate starts with the $819 difference in column 2; it is assumed, however, that productivity in the supplier sector is lower than the OEM level and that the wage ratio remains unchanged. As a first approximation the material advantage has been reduced to $700.

[c] Other costs include capital charges, general selling and administrative (GS&A) expenses, warranty costs, and costs of in-country transportation. The capital costs are taken from Table A.5 as described in the text; GS&A expenses have been taken from annual reports. The breakout is as follows: capital = $68; GS&A = ($132). The Japanese have an advantage in warranty of $90; their inland freight is $120 less expensive. These data are taken from a report by Harbour Associates as reported in Abernathy et al. (1980, p. 60).

[d] Numbers in parentheses indicate the U.S. advantage.

more accurate reflection of the situation. The calculations imply that Toyo Kogyo uses less labor and less capital per vehicle. While these data have not been adjusted for vertical integration or for differences in product mix, doing so would not affect the conclusion. If we use the same adjustment for product mix that we used in the labor productivity analysis (see Table A.3), we would reduce the Ford book value per vehicle from $3048 to $2340 (the adjustment factor is 0.768). If we also use the same adjustment for vertical integration that we used in the labor productivity analysis, we would increase Toyo Kogyo's book value per vehicle by 15 percent, from $1639 to $1885. The product mix adjustment is likely to understate capital per vehicle at Ford, because capital input does not increase with the size of the vehicle as rapidly as labor and materials.[20] Even so, these adjustments show that the Japanese firm uses less capital per vehicle than its U.S. competitor.

The evidence in this table reinforces the observation of the panel experts that differences in technology and automation, while factors, were not the most important determinants of differences in labor productivity. Moreover, the table adds further evidence that the overall cost and productivity gap lies in the more effective and efficient use of relatively comparable resources. If we use capital per vehicle adjusted for inflation and the specific exchange rate and assume a capital charge (depreciation plus interest) of 15 percent, the numbers in Table A.5 imply a Japanese advantage of $433 per vehicle in capital costs [($5052 - $2162) x 0.15]. If the Ford capital per vehicle is reduced to reflect differences in product mix as described above, and if the Toyo Kogyo capital per vehicle is increased by 15 percent to reflect differences in vertical integration, we end up with a Japanese advantage of $209 [($3880 - $2485) x 0.15]. Finally, if we use book value per vehicle and the general exchange rate, and apply the product mix adjustment and the adjustment for vertical integration, the result is a Japanese advantage of $68 [($2340 - $1885) x 0.15]. This is the value we have used in the overall cost comparisons. Because the product-mix adjustment may understate capital per vehicle at Ford, and because we have used the general exchange rate and book values, the $68 estimate may understate the Japanese advantage.

Summary

Table A.6 summarizes the cost differences obtained under the four approaches and adds estimates of other costs, indirect freight charges, and corporate overhead. These calculations are necessarily rough. And we have not maintained strict independence, since a given method may use information obtained under a different perspective. As noted throughout this appendix, where assumptions were necessary we have tried to err on the side of understating the cost advantage of the Japanese. Nevertheless, the results point to a significant differential ranging from $1000 to more than $1400.

NOTES

1. These estimates are based on unpublished work done at the Research Department of the United Auto Workers.
2. Baranson's estimates are presented in Toder (1978), p. 151.
3. British Central Policy Review Staff (1974).

4. These estimates are based on data from the Japanese Ministry of Labor and the U.S. Bureau of Labor Statistics.

5. This figure is based on unpublished data from the U.S. Bureau of Labor Statistics (BLS), as well as published company information (annual reports, etc.). Fringe benefits differ significantly between the U.S. and Japanese auto industries, and we have attempted to reflect these differences. Given the importance of "internal" fringes in Japan, there may be errors in the estimates. Discussions with knowledgeable participants in both the United States and Japan suggest that any errors are minor. For 1980 the BLS estimated the U.S. and Japanese compensation rates in the motor vehicles and equipment industry as follows:

	Average Hourly Earnings	Hourly Compensation
United States	$9.81	$14.71
Japan	$5.96	$ 6.98

Source: Unpublished data, U.S. Bureau of Labor Statistics, October 1981.

6. Note that the calculations are based on compensation and not the costs to the employer of a unit of labor. The ratio ignores costs to the employer for an hour of work that does not show up in the employees' direct compensation--taxes, absenteeism, and so on. Since these terms tend to be larger in the United States, the calculations may understate the Japanese advantage. It should be noted that the panel did not deal with the question of relative rates of compensation for top executives in the U.S. and Japanese auto industries. Although compensation data have been used for all salaried workers in the cost comparisons, no attempt was made to break out executive compensation (i.e., vice-president and above). The view that top executives in Japan have lower rates of pay than their U.S. counterparts has been expressed in the literature, but the panel had no basis for judging that claim. The absence of disclosure requirements and the complexity of top-executive pay make comparisons difficult. Furthermore, U.S.-Japan differences in top executive compensation are likely to have only minor effects on cost comparisons. For example, in a good year, such as 1978, the top five executives at Ford were paid an average of $638,000 in compensation. Assume for the moment that the top 25 executives at Ford were all paid this amount, while their counterparts at Toyo Kogyo were paid nothing. Such a difference in executive compensation would add only $6 to the per vehicle cost difference between Ford and Toyo Kogyo. (This assumes that all executive compensation applies only to U.S. passenger car production, which was about 2.6 million in 1978.)

7. Committee on Motor Vehicle Emissions (1974). Note that we have assumed an average level of options. Industry sources include staff reports of U.S. companies, memoranda from members of the panel, and informal discussions.

8. The nominal cost per hour worked is the cost to the employer and includes base rates, fringes, and other payroll costs. We assume that material suppliers (steel, plastics, etc.) have the same ratio of unit labor costs as participants in the industry.

9. Let C(US) and C(J) be unit labor costs in the United States and Japan, respectively. We estimate C(J)/C(US) = 0.425. We want to know C(US) - C(J). Column 6 gives us C(US). Thus, C(US) - C(UJ) = [1 - C(J)/C(US)] x column 6; this result is column 7.

10. See notes to Table A.6 for a breakdown. These estimates pertain to 1980; updating to 1981 would raise them somewhat.

11. This is documented in Abernathy et al. (1980).

12. The data for annual hours worked for Ford assume 35 days of contractual vacation, paid personal holidays, and so on and an adjustment for absenteeism of 5 percent of normal annual hours (assumed to be 2000). The Toyo Kogyo hours are based on 2037 hours average in Japanese manufacturing, plus adjustments based on holidays and vacations.

13. The vertical integration adjustment reflects industry judgement rather than analysis of purchases-to-sales ratios; on the latter basis the two companies are comparably integrated.

14. The procedure probably understates the number of hours per vehicle at Ford, because the adjustment is based on total manufacturing costs, not the cost of labor alone.

15. If applied to total passenger car production, the large cost difference reported here would seem to imply much higher profits than Japanese firms typically report. Since accounting practices differ between the two countries, it is difficult to interpret such apparent discrepancies. One possible explanation is the difference between profits for Japanese auto companies in their home market and profits in the U.S. market. Discussions with Japanese executives suggest that the U.S. market is much more profitable than the domestic market in Japan. The overall level of Japanese profits, therefore, may be an average of very profitable and only marginally profitable markets around the world.

16. If the industry data on productivity are accurate (Japan-to-U.S. ratio of about 1.2), and we use the Toyo Kogyo-Ford ratio at the OEM level (about 1.5), then the implied level of productivity for suppliers and the assumed wage ratio justifies a differential of about $650-$700 per vehicle.

17. Plant-level data on Japan are available from the Japan Ministry of Labor in its annual Labor Productivity Statistical Survey. U.S. data have been obtained from industry sources. For additional data, see Abernathy et al. (1980).

18. Abernathy et al. (1980).

19. See Abernathy (1978), pp. 86-113, for a discussion of technology in engine plants and pp. 114-146 for a discussion of assembly plant technology.

20. For evidence on this point, see Abernathy (1978), pp. 21 and 193-194, where the unit body construction method used on small cars is described. It appears that small-car production may actually be more capital intensive (more capital per vehicle) because of many fewer parts and unit construction.

Productivity and Absenteeism

This appendix examines the effects of unexpected absenteeism on redundant employment. As an explanation of productivity differentials, absenteeism may have two effects. First, where a pool of relief workers must be carried to cover for unexpected absences, some redundancy is likely to be experienced. The extent of this effect depends critically on the variation in the daily absenteeism rate. Second, fill-in workers may be less familiar with the absentees' jobs or not as effective in the affected work group. Given the nature of the jobs and the organization of work, however, industry sources generally view work disruption as a minor factor, with redundancy in the relief pool of much greater importance.

Table B.1 presents an analysis of absenteeism and productivity in the auto industry using data averaged over several firms and establishments. Before discussing the approach and results it must be stressed that we have focused on the impact of absence on labor hours per vehicle. Absenteeism will have an additional effect on costs through the fringe benefits that are paid to absentees. Even though straight-time wages are not paid to those absent, fringe benefits tend to be unrelated to hours worked and thus are paid irrespective of the number of days of absence. This effect has been captured in the employee cost per hour used in our earlier calculations.

The absenteeism analysis in Table B.1 assumes that only unexpected or unplanned absences are relevant to estimation of redundant labor hours. Time away from work that is predictable can be planned for so that no redundancy occurs. In the case of planned absence, only the effects of disruption in work groups or job unfamiliarity are relevant; we assume these effects to be relatively negligible. Using industry-wide data we estimate that unexpected, unplanned absenteeism averages 3-6 percent. The lower bound is obtained by counting only "absent without notice" as unplanned, while including all short-term absence yields the

189

TABLE B.1   Absenteeism and Productivity: U.S.-Japanese Differences

Category	United States[a]	Japan
Absenteeism (percentage of employed hours)		
Absent without notice	3.0	N/A
Medical	1.0	N/A
Personal	1.8	N/A
Other (jury duty, etc.)	0.6	N/A
Total	5.7	0.5-1.0
Redundant Labor Hours—Relief Pool		
Average unplanned absenteeism[b] (percentage)	4.5	0.75
Peak unplanned absenteeism[c] (percentage)	11.25	1.9
Average redundancy (percentage of employed hours)	6.75	1.15
Productivity Impact		
Labor hours per vehicle[d]	82.0	53.0
Absenteeism effect if U.S. redundancy drops to zero (hours per vehicle)	5.5	—
Absenteeism effect if U.S. redundancy drops to Japanese level (hours per vehicle)	4.6	—

[a] U.S. estimates are approximate industry averages based on data from panel members; Japanese estimates based on data from panel members.

[b] This assumes some of the medical, personal, and other absenteeism is planned.

[c] Assumes all absenteeism occurs on Monday and Friday; thus, 4.5 = (2/5 x%), where x% is the Monday (Friday) rate.

[d] For a small car, based on Ford and Toyo Kogyo estimates in Table A.3 in this volume.

upper bound; we use the midrange of 4.5 percent in subsequent calculations.

Even though the average is something like 4.5 percent, variation above the average may influence staffing decisions. To estimate an upper bound on the effect of absenteeism, we assume that all unplanned absence occurs on Monday and Friday. With no absenteeism in midweek, Monday and Friday will average 11.25 percent [(4.5 x 5)/0.2]. If we assume that the relief pool is staffed to the peak, then on average there will be 6.75 percent redundant hours of work (11.25 - 4.5). In other words, plants must hire 6.75 percent more labor hours than they actually need to produce a given level of output, simply to cover for unplanned absence.

What impact does this have on hours per vehicle? Using the Ford estimates for a small vehicle from Table A.3, the analysis implies that unplanned absence accounts for 5.5 hours per vehicle. This amounts to almost 20 percent of the estimated Ford-Toyo Kogyo productivity gap.

There are several reasons to suppose that 5.5 hours is an over-estimate of the true value. In the first place we have implicitly assumed that the relief pool must be hired for a full week, even though they work for only two days. The fact that some unplanned absence occurs on Tuesday, Wednesday, and Thursday would lower the redundancy estimate somewhat. Moreover, there is some evidence that production managers make use of short-term employees to cover 1- to 2-day shortages, without having to add them to the relief pool on the other days. There is also the obvious point that we have assumed that the Japanese producers have no unplanned absenteeism, when in fact industry sources suggest that the actual rate is likely to range from 0.5 to 1.0 percent. Applying the same analysis to the Japanese data yields a redundancy rate of 1.2 percent. If U.S. plants were to achieve that level, 4.6 hours or 16 percent of the productivity gap would be closed. Factoring in other adjustments probably reduces the effect to between 10 and 12 percent.

While not a dominant factor the analysis thus implies that absenteeism has a noticeable impact on the productivity differential. Clearly, when the effects of fringe benefits are added, its impact on overall costs could be sizeable. Industry sources suggest that from $100 to $150 in cost per vehicle could be eliminated with reductions in absenteeism to the Japanese level. Cumulated over several million vehicles, the absolute impact is sizeable.

Statistical Analysis of Technology, Sales, and Prices

The sales model underlying the estimated effects given in Table 8.3 is given by:

$$S_i - b_0 + \Sigma_i \beta_i X_{ij} + \epsilon_i \, ,$$

where S_i measures sales of the ith model, X_{ij} is the jth characteristic of the ith model, and β_j captures the effect of the jth characteristics on sales.

As specified here the equation is a reduced form. It is derived from a set of structural equations relating supply and demand to the price of a vehicle and its characteristics. Solving for price and sales in equilibrium yields the reduced form relating sales to characteristics.

The price equation is expressed similarly:

$$P_i = c_0 + \Sigma \lambda_j; X_{ij} + u_i \, ,$$

where P_i is the transaction price, u_i is an error term, X_{ij} is as defined as before, and λ_j is the effect of the jth characteristic on prices. The list price is added to the price model to give perspective on the importance of demand and supply effects.

The specification of the basic models is presented in Table C.1. As given there, many of the variables in the analysis are easily defined and readily available. For sales, prices, and age data, however, a number of assumptions were necessary. Price data for domestic cars were readily available only for models with automatic transmission, power steering, power brakes, and air conditioning. Prices of imported products, however, were readily available for models with standard equipment--generally a manual transmission, without the power accessories. We followed this convention in both list and first-year prices. Sales data were available only for the model line (e.g., Citation) and were not broken down by engine type or other features. Finally, model age

192

TABLE C.1 Basic Estimating Equations

Fully Specified Sales and Discount Equation (1977, 1979)

$$\frac{S_i}{P_i} = a_0 + a_1\,MPG_i + a_2\,RNG_i + a_3\,VOLWT_i + a_4\,REP1_i + a_5\,REP2_i + a_6\,DIESEL_i$$

$$+ a_7\,FWD_i + a_8\,AGE_i + a_9\,JAPAN + a_{10}\,JAPCAP + a_{11}\,USA + a_{12}\,SUBCOMP$$

was determined by using trade reports and expert commentary on new models to determine when a major redesign took place.

The results are presented in Tables C.3 and C.4. In the price results we find evidence of a change in the valuation of technology regardless of whether the list price is included. F-tests of the hypothesis that the λ_j is equal in the two years are reported in Table C.2. We see that once technology characteristics come into the picture we can reject the equality hypothesis. The sales results are much weaker. The large standard errors on the coefficients make clear that the equality hypothesis cannot be rejected. In general, however, the data provide little information about the effect of technology on sales.

TABLE C.2 Tests of Equality of the 1977-1979 Coefficients

Equation and degrees of freedom[a]	Calculated F-Statistic	Critical Ratio (95 percent confidence level)
1. (5,117)	1.37	2.67
2. (6,115)	2.18	2.52
3. (9,109)	3.88	2.23
4. (10,107)	3.33	2.17
5. (12,103)	2.31	2.07
6. (13,101)	2.08	2.03

[a] The equation numbers correspond to the equations in Table C.3 in this volume.

TABLE C.3 Estimation of Price Equations, 1977-1979 (standard errors in parentheses)

Year/Specification	CONS	SUBCOMP	MPG	RNG	VOLWT	DIESEL	FWD	AGE	P*	REP1	REP2	R²	SEE	d.f.
I. 1977 Results														
1.	4522.0 (654)	-1124.0 (286)	21.47 (31.4)											
2.	-132.4 (517)	-19.6 (164.3)	20.09 (15.06)						0.74 (0.0655)			.432	730.8	38
3.	7796.0 (1527.0)	-1015.0 (274.0)	69.4 (38.8)		-74.2 (34.0)	-18.3 (874.0)	-204.0 (421.9)	-33.1 (30.9)				.873	350.5	37
4.	165.0 (1166.0)	179.6 (172.9)	37.2 (21.1)		-7.8 (19.6)	-785.7 (477.1)	-148.1 (226.8)	-11.25 (16.8)	0.75 (0.08)			.593	653.6	34
5.	6909.0 (1849)	-827.1 (329.0)	23.7 (57.6)	3.8 (3.4)	-60.4 (38.5)	-186.0 (1070.0)	-251.0 (476.4)	-42.6 (37.6)		-55.0 (362.2)	-106.8 (616.7)	.886	351.3	33
6.	172.0 (1283.0)	-176.7 (195.7)	36.2 (31.6)	0.09 (1.94)	-8.1 (22.0)	-852.6 (1591.5)	-315.9 (261.5)	-11.3 (20.9)	0.74 (0.09)	26.0 (198.8)	64.7 (338.6)	.609	671.3	31
												.886	367.9	30

II. 1979
Results

Results														
1.	5814 (592)	−721.2 (195.5)	10.28 (21.63)									.289	727.5	79
2.	−553.1 (486.7)	226.1 (111.1)	31.55 (10.52)						0.80 (0.0497)			.837	351.1	78
3.	7516.0 (685.0)	−460.3 (186.9)	−55.6 (25.0)	28.9 (12.9)	1575.0 (426.0)	1417.0 (231.0)	−70.5 (30.0)					.576	576.7	75
4.	483.1 (639.4)	281.3 (115.1)	−0.86 (14.1)	−2.4 (7.2)	308.2 (248.9)	540.6 (140.9)	−35.1 (16.4)		0.71 (0.05)			.878	312.0	74
5.	6308.0 (766.0)	−370.7 (183.1)	−88.9 (26.8)	−9.7 (13.9)	1524.0 (410.0)	998.9 (264.2)	−51.5 (30.0)	3.9 (1.3)		126.6 (332.1)	19.2 (245.6)	.625	553.3	72
6.	311.3 (639.0)	312.8 (115.7)	−14.0 (16.1)	3.4 (7.8)	301.7 (248.0)	416.6 (154.6)	−36.8 (16.8)	1.0 (0.7)	0.70 (0.06)	74.4 (186.2)	221.9 (138.1)	.885	309.1	71

NOTE: The method of estimation was OLS; each equation includes dummy variables for country of origin.

TABLE C.4 Estimated Sales Equations, 1977-1979 (standard errors in parentheses)

Year/ Specification	CONS	SUB- COMP	MPG	RNG	VOLWT	DIESEL	FWD	AGE	REP1	REP2	R²	SEE	d.f.
I. 1977 Results													
1.	32.4 (78.7)	-38.1 (34.4)	0.9 (3.8)	—	—	—	—	—	—	—	0.233	88.0	38
2.	-129.2 (207.5)	-80.5 (41.7)	4.7 (7.2)	-0.7 (0.41)	5.2 (4.4)	78.3 (118.8)	-17.7 (41.5)	1.6 (4.2)	—	—	0.356	86.5	33
3.	-97.4 (212.0)	-87.8 (43.0)	6.6 (7.5)	-0.8 (0.4)	4.7 (4.5)	-0.6 (141.3)	-23.4 (42.3)	1.4 (4.8)	7.9 (46.5)	82.2 (74.5)	0.381	87.5	31
II. 1979 Results													
1.	-29.2 (77.2)	-12.6 (32.2)	2.9 (2.9)	—	—	—	—	—	—	—	0.224	88.8	48
2.	-174.4 (163.8)	-1.7 (3.7)	2.5 (5.0)	-0.1 (0.3)	3.8 (2.6)	-70.1 (94.5)	-29.1 (48.8)	-0.3 (7.1)	—	—	0.318	87.8	43
3.	-189.1 (161.3)	3.0 (36.4)	2.1 (4.9)	-0.1 (0.3)	4.2 (2.5)	-97.8 (94.0)	-53.1 (49.6)	-0.8 (7.1)	-61.3 (65.0)	65.2 (39.8)	0.372	86.4	41

NOTE: The method of estimation was OLS; each equation includes dummy variables for the country of origin.

Bibliography

Abernathy, William J. "Innovation and the Regulatory Paradox:
    Toward a Theory of Thin Markets," pp. 38-64 in Ginsburg and
    Abernathy, eds., Government, Technology and the Future of
    the Automobile (New York: McGraw-Hill, 1980).
Abernathy, William J. The Productivity Dilemma (Baltimore,
    Md.: The Johns Hopkins Press, 1978).
Abernathy, William J., and James M. Utterback. "Patterns of
    Industrial Innovation," Technology Review (June-July 1978), pp.
    40-47.
Abernathy, William J., and Kenneth Wayne. "Limits of the
    Learning Curve," Harvard Business Review (Nov.-Dec. 1974),
    pp. 109-119.
Abernathy, William J., and Phillip L. Townsend. "Technology,
    Productivity and Process Change," Technological Forecasting
    and Social Change (Aug. 1975), pp. 379-396.
Abernathy, William J., James Harbour, and Jay Henn.
    "Productivity and Comparative Cost Advantages in the U.S.
    and Japanese Auto Industries." (Report to the U.S.
    Department of Transportation) (Dec. 1980).
British Central Policy Review Staff. The Future of the British
    Car Industry (London: Her Majesty's Stationery Office, 1974).
Byron, George E. Facilities Planning and Capital Investment (U.S.
    Department of Transportation-TSC) (April 1980).
Chandler, Alfred D., Jr. Giant Enterprise: Ford, General Motors
    and the Automobile Industry (New York: Harcourt, Brace and
    World, 1964).
Committee on Motor Vehicle Emissions, National Research
    Council Consultants Report, Manufacturability and Costs of
    Proposed Low Emissions Automotive Engine Systems
    (Washington, D.C.: National Academy of Sciences, 1974).
Doz, Yves. L. "Competition in Worldwide Industries. The
    Automobile Industry" (unpublished paper, 1979).

197

Fine, Sidney. Sit Down! The General Motors' Strike of 1936-1937
(Ann Arbor: University of Michigan Press, 1969).

Flink, James J. America Adopts the Automobile, 1895-1910
(Cambridge: MIT Press, 1970).

Flink, James J. The Car Culture (Cambridge: MIT Press, 1975).

Fuller, Mark B. "Note on the World Auto Industry in Transition"
(Harvard Business School Case Services, 1981).

Goodson, R. E. Federal Regulation of Motor Vehicles: A Summary
and Analysis. (Report to the U.S. Department of
Transportation) (Washington, D.C.: U.S. Department of
Transportation, 1977).

Guest, Robert H. "The Man on the Assembly Line: A Generation
Later," Tuck Today (May 1973), pp. 1-8.

Guest, Robert H. "Quality of Work Life--Learning From
Tarrytown," Harvard Business Review (July-Aug. 1979), pp.
76-87.

Ginsburg, D. H., and W. J. Abernathy, eds. Government,
Technology and the Future of the Automobile (New York:
McGraw-Hill, 1980).

Halpern, Paul J. Consumer Politics and Corporate Behavior: The
Case of Automobile Safety (Ph.D. dissertation, Harvard
University, 1972).

Hanson, Kirk O. "The Effect of Fuel Economy Standards on
Corporate Strategy in the Automobile Industry," pp. 144-161 in
Ginsburg and Abernathy, eds., Government, Technology and the
Future of the Automobile (New York: McGraw-Hill, 1980).

Hayes, Robert H. "Why Japanese Factories Work," Harvard
Business Review (July-Aug. 1981), p. 56.

Heywood, John, and John Wilkes, "Is There a Better Automobile
Engine?" Technology Review (Nov.-Dec. 1980), pp. 18-29.

Japan Ministry of Labor. Year Book of Labor Statistics (1979).

Japan Ministry of Labor. Annual Labor Productivity Statistical
Survey (1980).

John, Richard, et al. "Mandated Fuel Economy Standards as a
Strategy for Improving Motor Vehicle Fuel Economy," in
Ginsberg and Abernathy, eds., Government, Technology and the
Future of the Automobile (New York: McGraw-Hill, 1980).

Katz, Abraham. Statement of Abraham Katz, Assistant Secretary
of Commerce for International Economic Policy, before the
Subcommittee on Trade of the House Ways and Means
Committee. March 18, 1980.

Klein, Burton H. Dynamic Economics (Cambridge: Harvard
University Press, 1977).

Krugman, Paul. "Scale Economics, Product Differentiation, and
the Pattern of Trade," American Economic Review (Dec.
1980), pp. 950-959.

Lancaster, Kelvin J. "Intra-industry Trade under Perfect
Monopolistic Competition," Journal of International Economics
(Vol. 10, 1980), pp. 15-175.

Leone, Robert, et al. Regulation and Technological Innovation in the Automobile Industry (Report for Office of Technology Assessment, July 1981).

Mills, D. Q. "The Techniques of Automotive Regulation: Performance versus Design Standards," pp. 64-76 in Ginsburg and Abernathy, eds., Government, Technology and the Future of the Automobile (New York: McGraw-Hill, 1980).

Monden, Yasuhiro. "What Makes the Toyota Production System Really Tick?" Industrial Engineering (Jan. 1981), pp. 37-46.

Nader, Ralph. Unsafe at Any Speed: The Designed-in Damage of the American Automobile (New York: Grossman, 1965).

Nevins, Allan, and Frank E. Hill. Ford: Decline and Rebirth (New York: Scribners, 1962).

Nevins, Allan, and Frank E. Hill. Ford: The Times, the Man, the Company (New York: Scribners, 1954).

Rae, John B. American Automobile Manufacturers: The First Forty Years (New York: Chilton, 1959).

Rae, John B. The American Automobile: A Brief History (Chicago: University of Chicago Press, 1965).

Reuther, Victor G. The Brothers Reuther and the Story of the UAW: A Memoir (Boston: Houghton Mifflin, 1976).

Ronan, Lawrence, and William J. Abernathy. "The Development and Introduction of the Automobile Turbocharger." (Harvard Business School Working Paper #78:43, 1978a).

Ronan, Lawrence, and William J. Abernathy. "The Honda Motor Company's CVCC Engine: A Case Study of Innovation." (Harvard Business School Working Paper #78:43, 1978b).

Salter, Malcolm, and Mark B. Fuller. Profile of the World Automotive Industry (unpublished report, 1980).

Sloan, Alfred P. My Years with General Motors (New York: Doubleday Anchor, 1972 edition).

Sorensen, Charles E. My Forty Years with Ford (New York: W. W. Norton, 1956).

Stobaugh, Robert, and Daniel Yergin. "After the Second Shock: Pragmatic-Energy Strategies," Foreign Affairs (Spring 1979), pp. 836-871.

Stobaugh, Robert, and Daniel Yergin. "Energy: An Emergency Telescoped," Foreign Affairs (Dec. 1979), pp. 536-595.

Stockman, David. "The Wrong War? The Case Against National Energy Policy," Public Interest (Fall 1978), pp. 3-44.

Toder, Eric. Trade Policy and the U.S. Automobile Industry (New York: Praeger, 1978).

Toyo Kogyo Company. Annual Report (various issues).

Toyoto Motor Company. Annual Report (various issues).

Tracy, Karen. "Note on Automobile Emissions and Fuel Economy Regulation" (unpublished report, 1978a).

Tracy, Karen. "Note on Automobile Safety Regulation" (unpublished report, 1978b).

U.S. Department of Labor. "The General Motors Corporation Strike," Monthly Labor Review (March 1937).

Utterback, James M. "Innovation in Industry and the Diffusion of Technology," Science (Feb. 1974), pp. 620-626.

Utterback, James M., and W. J. Abernathy. "A Dynamic Model of Process and Product Innovation." Omega (Vol. 3, No. 6., 1975), pp. 639-656.

Vernon, Ray. Storm Over the Multinationals: The Real Issues (Cambridge: Harvard University Press, 1977).

Wells, Louis T., Jr. "The International Product Life Cycle and United States Regulation of the Automobile Industry," pp. 270-305 in Ginsburg and Abernathy, eds., Government, Technology and the Future of the Automobile (New York: McGraw-Hill, 1980).

White, George. "Management Criteria for Effective Innovation," Technology Review (Feb. 1978), pp. 14-23.

White, Lawrence J. The Automobile Industry Since 1945 (Cambridge: Harvard University Press, 1971).

Whitman, Marina V. N. "International Trade and Investment: Two Perspectives" (Graham Memorial Lecture, Princeton University, Department of Economics, 1981).

Wilkins, Mira. "Multinational Automobile Enterprises and Regulation: An Historical Overview," in Ginsburg and Abernathy, eds., Government, Technology and the Future of the Automobile (New York: McGraw-Hill, 1980).

Biographical Sketches

WILLIAM J. ABERNATHY has been a professor of business administration at the Harvard University Graduate School of Business Administration since 1972. Before joining Harvard he taught at Stanford University and at the University of California at Los Angeles. Dr. Abernathy holds a B.S. in electrical engineering from the University of Tennessee and an M.B.A. and D.B.A. from Harvard University.

ALAN A. ALTSHULER is a professor in and the chairman of the Political Science Department at the Massachusetts Institute of Technology. He received a B.A. from Cornell University in 1957 and an M.A. and Ph.D. from the University of Chicago in 1959 and 1961, respectively. Dr. Altshuler has taught at several institutions, including Swarthmore College and Cornell University, and has served as the chairman of the Governor of Massachusetts' Task Force on Transportation. He is a member of the National Academy of Public Administration and the American Political Science Association and is the author of several books.

JAMES K. BAKKEN has been Vice-President, Operations Support Staff, at the Ford Motor Company since 1979. Before assuming that position he was Vice-President, Body and Assembly Operations, at Ford. Mr. Bakken holds a degree in engineering from the University of Wisconsin and an M.B.A. in industrial management from the Massachusetts Institute of Technology. He has held numerous positions at Ford, having begun his career there as a student engineer in 1945.

KIM B. CLARK is an assistant professor at the Harvard University Graduate School of Business Administration. He holds an A.B., an A.M., and a Ph.D. in economics from Harvard University and is a member of Phi Beta Kappa. Before joining the Harvard faculty, Dr. Clark worked as an economist in the Office

201

of the Secretary, U.S. Department of Labor. He is the author of numerous papers and has served as a member of The Brookings Institute's Panel on Economic Activity.

DONALD F. EPHLIN is Vice-President and Director of the National Ford Department of the United Auto Workers (UAW). Before assuming that position, he served as the director of the UAW's Region 9A and as administrative assistant to the then UAW President, Leonard Woodcock. Mr. Ephlin began his career with General Motors and has spent most of his career with the UAW, having served as president of one of its locals from 1949 to 1960.

DONALD A. HURTER is the manager of Arthur D. Little's Automotive Technology Unit. Prior to joining Arthur D. Little, Mr. Hurter served in a managerial capacity with several firms, including six years as Vice-President, General Manager, of the Aircraft Division of Standard Thomson Corporation. He received his B.S. in mechanical engineering from the Massachusetts Institute of Technology and his M.S. in engineering from Yale in thermodynamics and internal combustion engines. He is a member of the Society of Automotive Engineers and is a registered professional engineer in the Commonwealth of Massachusetts.

TREVOR O. JONES, Vice-President and General Manager, Transportation and Electrical and Electronics Operations, TRW, Inc., is responsible for all engineering activities within that organization. Prior to joining TRW in June 1978, Mr. Jones spent 19 years with General Motors (GM). His most recent position was as Director of GM's Proving Grounds, which he assumed in 1974. Mr. Jones served on the National Motor Vehicle Safety Advisory Council, U.S. Secretary of Transportation, in 1971 and was appointed Vice-Chairman of the council in 1972. In 1975, President Ford appointed him to a three-year term on the National Highway Safety Advisory Committee. He received the U.S. Department of Transportation's Safety Award for Engineering Excellence in 1978. He is a fellow of the British Institute of Electrical Engineers and received its Hooper Memorial Prize in 1950. He is also a fellow of the American Institute of Electrical and Electronics Engineers. Mr. Jones holds many patents and is the author of many papers on the subjects of automotive safety and electronics.

HELEN R. KAHN is the Washington Bureau Chief for Automotive News. She holds a B.A. degree, magna cum laude, from Bridgewater College and an M.A. from Vanderbilt University. From 1944 to 1947 she worked for <u>Pathfinder Magazine</u> and for the U.S. Army Intelligence doing crytoanalysis for the Bureau of National Affairs. She has taught in the English Department of the University of Maryland and has been a free-lance writer of magazine articles and ghost writer for books.

DUANE F. MILLER is Vice-President, Engineering, at Volkswagen (VW) of America, Inc. His entire professional career prior to joining VW in 1977 was at the Pontiac Motor Division of General Motors. Mr. Miller holds a B.S.M.E. from the University of Nebraska and an M.B.A. from Michigan State University. He is a member of the Society of Automotive Engineers, the Engineering Society of Detroit, and Rotary International.

RALPH L. MILLER is Vice-President of Market Development and Strategic Planning for the Light Vehicles Group at Rockwell International. Prior to assuming that position, Dr. Miller was the Director of Manufacturing Facilities Planning, Worldwide Product Planning, at General Motors. He has a B.A. degree from Amherst College and a Ph.D. from the Massachusetts Institute of Technology. Dr. Miller has taught at the University of Detroit and was marketing manager for Adam Opel in Germany.

RICHARD H. SHACKSON is the President of Shackson Associates, Inc. Prior to founding his own business, Mr. Shackson was Assistant Director of Transportation Programs at the Energy Productivity Center, Carnegie-Mellon Institute of Research, and the Director of Environmental Research at the Ford Motor Company. He holds a B.S.E.E. from the Case Institute of Technology and has served on the National Academy of Sciences' Transportation Research Board, Office of Technology Assessment Advisory Panels, and has served as President of the Transportation Research Forum.

PETER D. ZAGLIO is Vice-President, Securities Division, Lehman Brothers Kuhn Loeb. His professional analytical career, which has included employment at Smith Barney; Harris Upham; and Loeb Rhoads, Hornblower, has been focused on the automotive industry. He holds a B.A. in economics from Cornell University and an M.B.A. in finance from the Wharton Graduate School.